Making A Living Alongshore

Making A Living Alongshore

by Phil Schwind

Illustrated by
Coralee Spacht Hays

International Marine Publishing Company
Camden, Maine

International Standard Book Number 0-87742-080-7
Library of Congress Catalog Card Number 76-8785

CONTENTS

INTRODUCTION

With the emphasis on Norwegian sardines, Icelandic haddock, Canadian salt codfish, South African lobster tails, Japanese crabs, and so on and so on, what has happened to the American "alongshore fisherman," the little fellow who provided his neighborhood with fish and shellfish that were fresh, *fresh,* FRESH, right out of the water with their tails still wagging, or with their shells clamped shut? The fish and shellfish are still there, but the fine art of making a living along the shore has gone the way of blacksmithing and candle making. Mass production may have done away with the handcrafts, but the fish and shellfish are still there, waiting to be caught. They may exist in smaller numbers because of competition, but they are more valuable because of higher prices, and they are more vulnerable because of better equipment.

The often lonely way of the alongshore fisherman is not for everyone, but for the right person there is still a living to be made almost everywhere in the salt marshes, in the estuaries. It is a trade without glamour, a trade that takes a great deal of ingenuity, but it is a way of living, not a way of life, one without parallel in freedom and personal satisfaction.

The tools are many and varied and many are handcrafted. I will try to describe them in enough detail so you can reproduce them. More than that, I will try to explain *why* they are made the way they are, so that an understanding of the tools may help you make ones that are better and more efficient.

The blind, dumb attitude, "That isn't the way my father taught me," or, "That isn't the way the book says," has always driven me to extreme frustration. I've lost a couple of otherwise very special partners because I couldn't stand it. Probably my attitude of, "Let's try something new," or, "There must be a better way," was equally upsetting to those same partners. You *do* have to start off with a little of, "This is the way it has always been done and there's a reason for it," but there is always a better way. While the search for it can be expensive, sometimes almost catastrophic, the satisfaction and the profit of finding a better way can be worth all the trials.

Much of what follows is open to criticism. There will be a great deal of, "That isn't the way it's done locally," or, "That isn't the way I learned it." So?

All of the methods and all of the gear hereinafter described worked well for me in the areas I was fishing. Much is traditional, handed down from the Ancients. (Since I am in the way of being an old-timer, those who taught me must be ancient.) But some of this information is new, and I have no doubt that there are even newer, better methods, and newer, better materials.

I learned the traditional ways from a dozen friends, but since most of them have gone on to better fishing grounds and their names would mean nothing to you, your not knowing them is your loss, my knowing them was my good fortune. However, I can pass on to you some of the bits of knowledge they passed on to me—some techniques, some "wrinkles"—and thus in a small way perhaps I can repay them. Being the men they were, they would surely want it that way.

Not all the know-how was acquired the easy way. Some of it was by a sort of osmosis: I sort of soaked it in. My good wife had justification when I came home after a blowy day ashore, to greet me with, "Where have you been? You promised to put up the storm sash . . . ," or whatever. But it was by listening to stories of the same fish being caught over and over that I learned little by little to try this or forget that. That and a continual dissatisfaction with the way it was done. Much of what is here was learned with cut-and-try, much was make-do. No government committees investigated us, we had no federal subsidies. If we weren't making enough to pay the grocery bill, we changed, and changed, and changed again until whatever gear we were using worked—or we had to give up for awhile and go to something else to make a dollar.

There's one thing that bothers me about the present generation of fishermen: they act as though they were either lonesome or timid. They act as though whatever the crowd does is the best way. Nobody goes "looking" any more; nobody seems to want a better way or a hotter place. It's always the itchy character who wants to see what's on the other side of the next wave who finds a new spot, who invents new gear, who pulls the whole fleet in his wake.

I learned one other principle from the Ancients: "Get in when the band begins to play and get out before the coda."

Let's take an example: We dug clams until Labor Day. After that date, that year's supply of clams was thinned down, and soon the market was even thinner. But littlenecks were up, and one canner or another wanted big quahogs. That was only a stopgap because, come the first of October, the scallop season opened. Then everybody and his dog and his grandmother's cousin got into the scallop business and the price dropped proportionately. The first of November opened the flounder dragging season. That took a certain amount of gear, so not everyone could get into the flounder business. We dragged them, first in one salt pond and then in another, finding that if we went back in a couple of weeks we caught almost as many the second time around as we had the first. And then we got frozen in, good and solid. So what did we do? We speared eels through the ice. If we had a January thaw, we either went back to scallop dragging or, if there were any left and the price was right, we hit the flounders again. Spring saw the handliners wanting moonsnails for bait (sometimes we made more money than the handliners), and by May the stripers were in the marsh. If striper fishing was thin when the moonsnail market eased off (for years we forgot clams because "there weren't none," though that's not the case now), then we ourselves went handlining for flounders, but we were geared for stripers if they struck in again. We ran conk pots some years and eel pots others.

As you can see, there was a continual change as the market moved up or down. We followed no man's compass course and fished in no one's wake. Yes, it took a lot of gear and a lot of know-how—that's the reason for this book. No one can tell you where, and when, and how to fish, but if you'll sit up and listen a little, if you'll realize there are other ways to make a living than hanging on the end of a quahog pole, or going on welfare, you may not make more money, but you'll have a lot of fun, and you won't have to wear any man's tie or take off your hat to any damned man. We escaped the city during the "Great Depression." You might just be headed that way. We found that if we couldn't always sell what we caught, at least we could eat it. It might help to keep that in mind.

There might seem to be two glaring omissions in this volume. The first is a discussion of how to dig sea worms for market. Since this is an occupation confined largely to Maine, my suggestion is that you can get far more expert advice from local fishermen than I can give, once you have proved to them that you are making an effort. The second omission is the process of rafting oysters in Connecticut, and again my suggestion is that you consult the professionals if you are considering this as a way of making a living. Every man to his trade, as my granddaddy often said, and a shoemaker should stick to his last—though he didn't say last what.

While I may seem facetious at times, the information herein is honest and sincerely intended to be helpful. I have included nothing that does not have experience to prove it and back it up. That there are many other ways to make a living alongshore, I have no doubt. These ways I have tried. May my experience help solve some of your problems.

1 COMMERCIAL SURF FISHING

There are a dozen or more books on the shelves (including my own: *Striped Bass and Other Cape Cod Fish,* Chatham Press, 1972) on how to catch stripers in the surf, but none of them, as far as I know, deals with how to make money at the business. Most of the books are full of tips on where to go, and when and what lures to use. All of them have valuable information, but sportfishing is not and should not be the same occupation as fishing for a living.

"Meat fishing" has only one valid formula: money made minus money spent equals profit. No other consideration is pertinent. If the time involved matters to you, get a nine-to-five job, pay your withholding tax, and go surf fishing for the fun of it.

You will have to know at what time you caught fish this morning in order to know when you should be at the beach tomorrow, but the fish don't know. No fish wears a wristwatch. The fisherman who goes out earliest, works hardest, and stays longest (and this holds true for any kind of fishing) will eventually catch the most fish and thereby make the most money—luck, skill, and the weather notwithstanding. If you and I and another fisherman go one day, and one of us comes in with the most fish, that's probably luck. But if one of us comes in consistently high line, day after day, that's not luck. That's hard work and long hours. Furthermore, it's fine to share your fishing with your buddies, and that's the way it should be if you're fishing for fun, but if you're

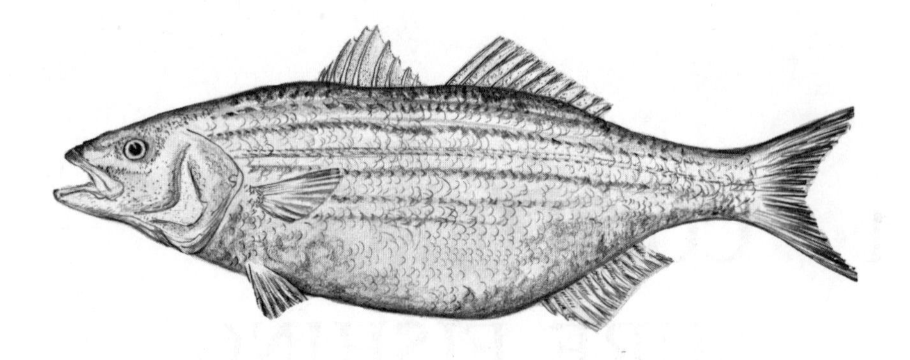

Striped bass, *Roccus saxatilis,* the money fish in the surf.

fishing for market, keep your mouth shut. If you have to brag, wait until next winter, when the fish have gone.

As far as tackle goes, I doubt that the average commercial or moonlighting fisherman realizes how much he owes to the sportfisherman. Rods, reels, line, artificial lures—almost all of them were developed by sportfishermen who had the time and got some of their kicks by finding a new way to fish. Bait fishing? No. Natural bait and the techniques of using it are largely commercial, although a great many sports use bait. Which is not to say commercial fishermen should stay away from artificials. Don't get in a rut; be flexible. So much for generalities; let's get down to specifics.

RODS

You are going to need at least one good rod, and probably more. Let's face it: you can't use the same rod to bang out a six-ounce pyramid sinker and two herring heads for bait that you would need to work successfully a 1½-ounce top-of-water plug. No such rod has ever been made. And if you're going to do any considerable amount of skiff fishing in tidal estuaries, a 10-foot surf rod can be a mighty clumsy piece of equipment.

Ideally, then, you should start off with a fairly stiff, 10-foot, one-piece rod (fiberglass, of course). The two-piece rods are handier for carrying, I'll concede, but the joint will "click" and delay action slightly, and this will cut down on your distance. Furthermore, the balance will be slightly off. Twelve-foot rods are for the self-styled, bragging, "high-surfers." While landing a 40-pounder in high surf may be exciting, this kind of fishing will put little jam on your bread. You'd do better to spend your time in some sheltered estuary, fishing for "schoolies." Unless you have the build and the weight of a pro football player, you won't get all the distance a 10-foot rod has to give, let alone anything more.

I like to start with my own glass blank. A Carboloy tip, the best money can buy, saves money, line, and fish in the end (no pun intended). Three or four graduated guides are placed proportionately to the taper of the rod, and my own built-up reelseat is placed a little farther up on the rod than is standard, to give more leverage. A lighter rod for small artificials should be tied the same way or, if it is a foot shorter, it should be tied in the same proportions.

Boat rods are a different tool, bearing the same relationship to surf rods that a hatchet does to an ax. Here the purpose is not to get your bait or lure as far away from you as you can cast, but simply to get your line away from your boat. I like a five-foot tip and a 14-inch butt, with the reel seat an integral part of the butt. The rod will work better if it has a roller tip, whether you use Dacron or lead-core line for trolling. If you are going to use Monel for trolling, you must have roller guides, so while you're at it, equip your trolling rods with roller guides and tips. If you're trolling artificial baits, you'll probably want two rods (while you're at it, be sure your rod holders are deep enough and secured well to the gunwales of your skiff).

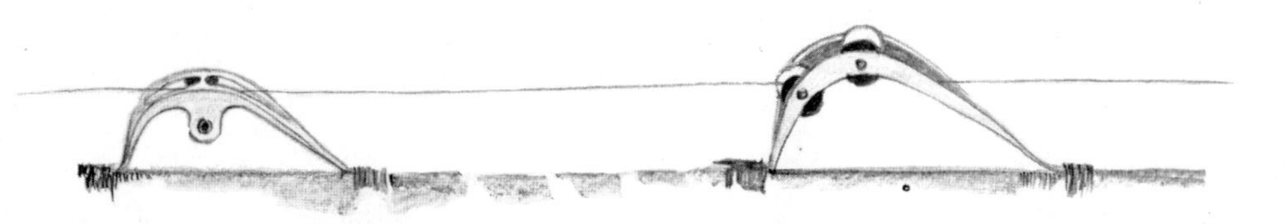

Roller guides are a must for wire line trolling.

REELS

If your reels are going to do double duty, for casting and trolling, casting reels will work for trolling, but trolling reels won't do for casting. Stick with wide-spool "flat" reels if you are buying "free-spool" tackle. If you are going all out for spinning, balance your reels to your casting rods (it won't make as much difference when you are trolling). If you have a heavy-duty stick, something with which you can cast out a heavy sinker and cumbersome baits, then the biggest of spinning reels is not too big. But if you have a lighter, shorter rod with a buggy-whip tip, cut down on the size of your reel. Use a middle-size or even a small reel, whichever feels better when you wield the rod.

To go back to trolling free-spool reels: they are actually better designed for this work than spinning reels, which take considerable expertise in letting your line run. If you're going to troll with deep line, whether lead core or Monel, you won't want a casting reel; you'll want a narrower, deep-spool reel with a faster take-up or retrieve ratio.

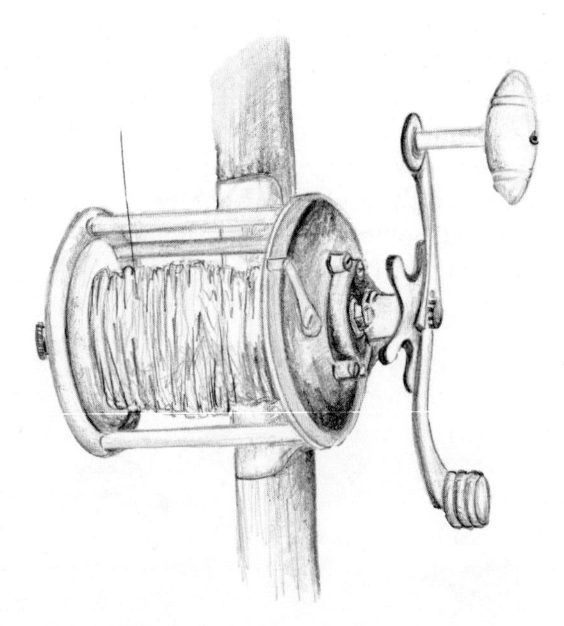

If you are using "free spool" tackle, stick with the wide-spool reel.

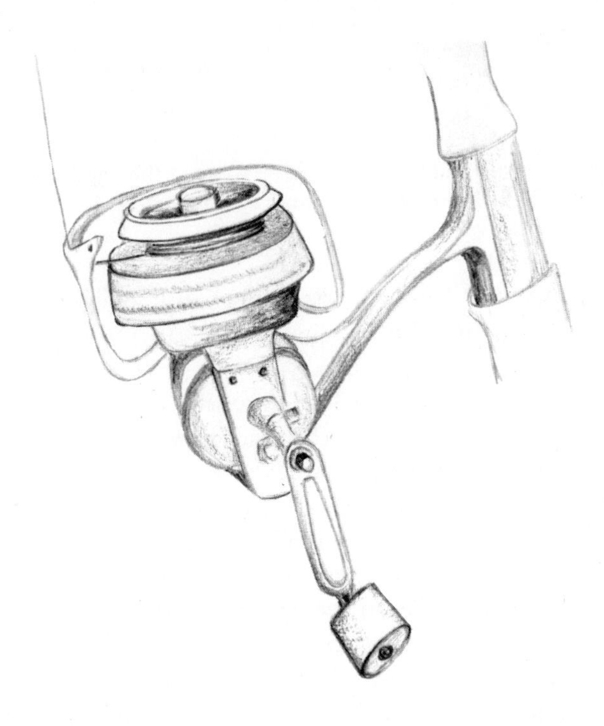

Balance your spinning reel to your rod: use the biggest reel for a heavy rod, a lighter reel for a slimmer rod.

LINE

If you plan to cast with a free-spool reel, heavy gear, 36-pound, braided nylon casting or "squidding" line is ideal. Keep your reel full of line—even if you have to put backing line under your working line—but not so full that the line binds on the crossbars. If you plan to cast with heavy spinning tackle, 30-pound monofilament is heavy enough, but get the limpest line you can buy. The cheaper, stiffer mono will give you trouble by "bird's-nesting" after very little use. Either type of line will do for trolling. If you have free-spool reels used only for trolling with heavy gear, 40-pound Dacron will probably last as long as you want to fish. (It won't do for casting, because it absorbs no water and can cauterize your right thumb more quickly than a hot stove.) Light spinning tackle for trolling, of course, means lighter monofilament, but if you are

On light spinning tackle, add an extra guide to keep the line from straying. The first guide should be as big as the face of the spinning reel.

fishing for market, 12-pound test is probably as light as makes any sense. Since the lighter line is finer, you can put more of it on a smaller reel.

If you plan to troll with lead-core line (which, while it is very popular, I do not like, because the lead core tends to corrode inside the nylon covering and thereby weaken the line) or Monel (which is nasty stuff to use but fishes very deep), 40-pound-test wire is usually strong enough. If, however, you plan to use a "Christmas tree" or "coat-hanger" multiple-rig lure, you may want 100-pound wire. You'll probably have to learn the hard way how to fill your wire-line reel. It is common practice to use 100 feet of wire tied to the Dacron, which fills the spool, but I find it more economical in the long run—and lose far less fish with lines that break at the knot—to fill the spool with wire all the way, even though it may take 200 yards of wire. If you *do* use Monel, use the solid rather than the braided; one broken strand in the braided line makes a thumb-snatcher. I'm sorry to have to say it, but I've never found a good soft-drawn stainless line that can compare with Monel, or even with lead-core line.

END TACKLE

Keep it simple. Leather thongs, three-way swivels, and safety pins are out. Not only do these gadgets cost money, but any manmade tackle can and eventually will part and/or snarl, and they will add nothing to your catch. Unless the fishing is very hot, I think almost everyone uses two hooks when surf fishing on the bottom with bait.

There are two schools of thought on the size of hooks. One insists that you cannot get enough bait on a small hook and so uses hooks as big as 10/0; the other feels that, since stripers tend to "mouth" the bait before they take off, a small hook, such as a 6/0,

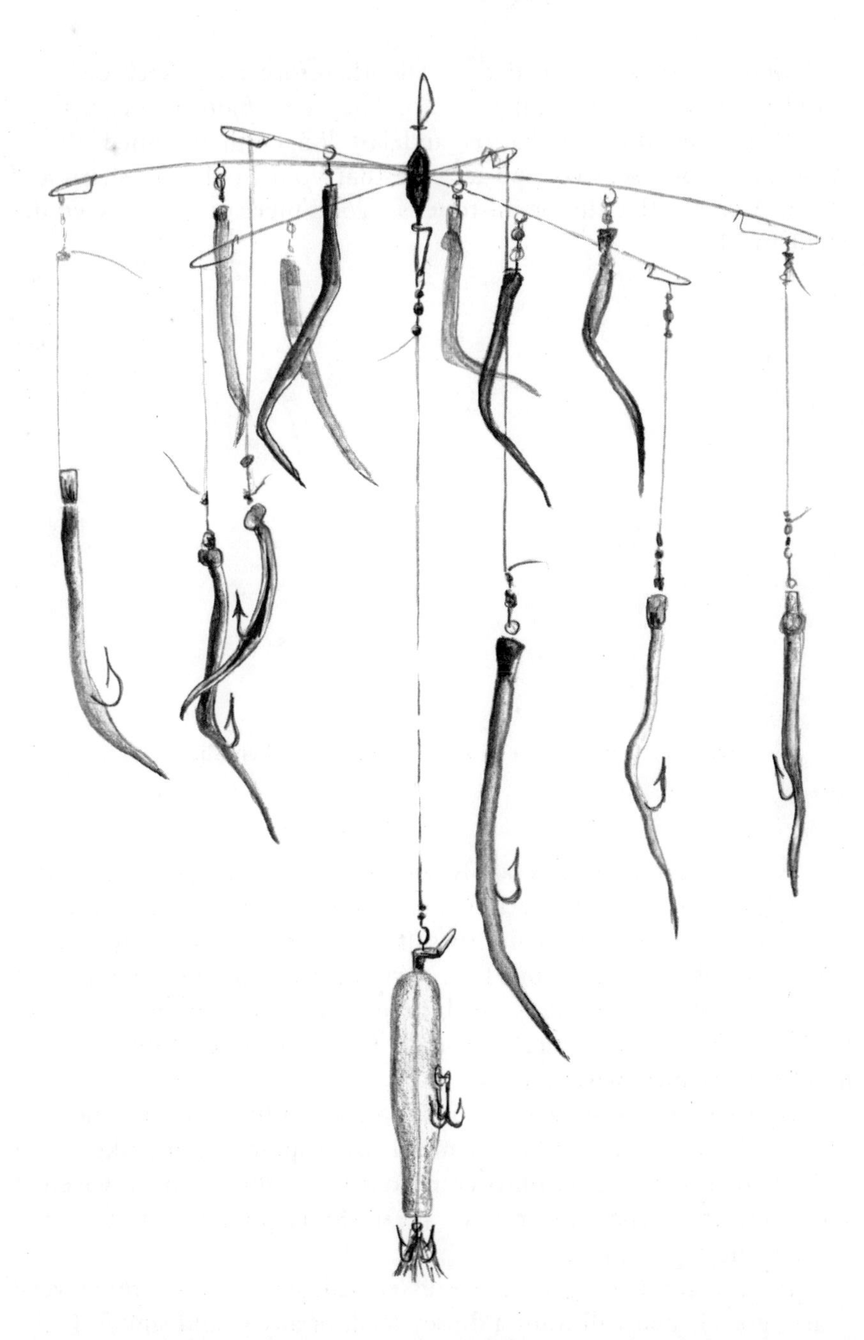

The "coat-hanger" or "Christmas tree" rig. While it is not everyone's favorite, it catches stripers.

will hook a striper more deeply and therefore more securely. Use stainless hooks if you want to. I do. They cost more to begin with, but they generally are sharper and last longer than tinned hooks. Whichever you use, take special care that your hooks are sharp and clean. A small, flat file or oilstone is a good piece of auxiliary equipment for this.

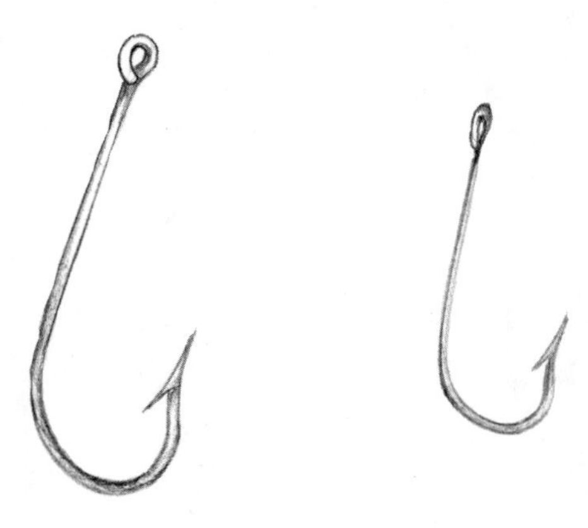

Whether you use large or small hooks, keep them sharp.

A leader does not necessarily serve the same purpose in saltwater fishing as it does in fresh water, i.e., to disguise the fact that there is a line attached to your bait. In salt water it acts as a handle with which to boat or beach your fish. A good, stiff, 30-pound mono, not more than two feet long, is ideal, unless there are bluefish or weakfish (that is, fish with teeth) around. In that case, use number 7 or number 9 stainless wire leaders.

A swivel does not swivel; it is simply a handy means of attaching your leader to your line (and a means of stopping your sinker from sliding down your leader onto your hook). Get the stainless swivels if you can; they last longer than brass. Sixty-pound test is strong enough and big enough.

If you are fishing over a mussel bed, or in rocks, or in very coarse gravel, you will want a dipsey lead, or any round sinker. If the bottom is clear sand or mud, a pyramid sinker will hold amazingly well in all but the heaviest surf. Whichever you use, the sinker should

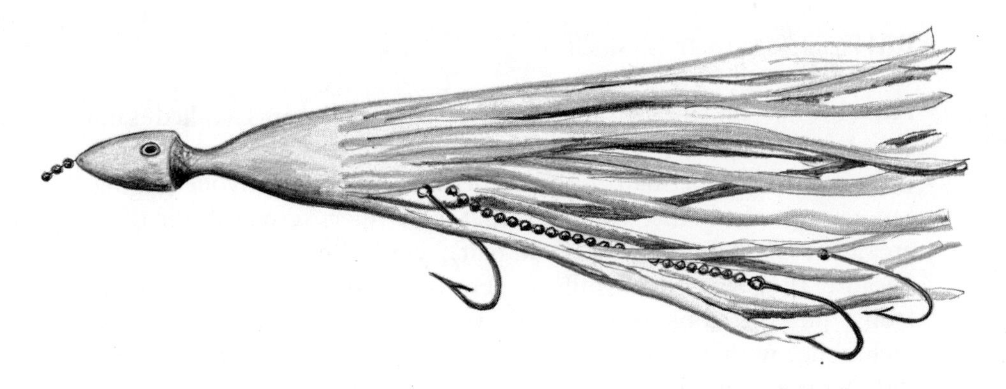

The "hootchie" is currently a very popular striper and bluefish rig for fast trolling.

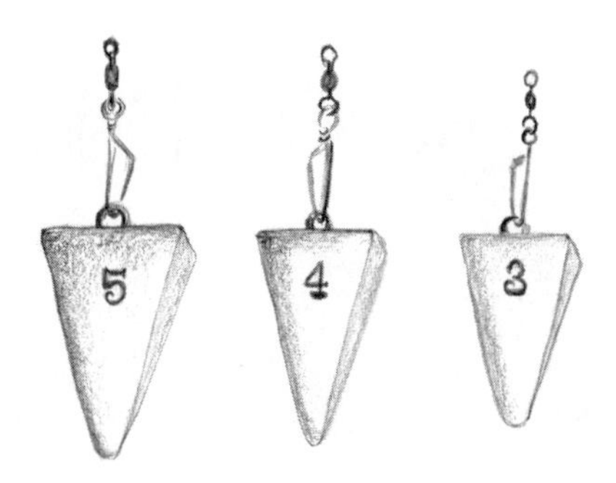

Balance a pyramid sinker to the gear, the surf, and the bottom.

be rigged so it will slide up your line when your fish starts to take off, thus giving him no leverage when his tail bangs against your line. You can wire a sinker to a swivel eye or you can use a stainless safety-pin swivel for the same purpose. Or you can use a "fish-finder," although I find the nylon type too bulky and complicated, and thereby prone to tangle.

Of the pyramid sinkers, a four-ounce sinker is about standard for heavy fishing. You may have to go to a heavier weight than that if you fish high surf or if there is an excessive amount of run-back to the surf. If you're fishing lighter gear, a three-ounce sinker probably is heavy enough. The lighter you fish, the more sensitive will be your touch, as long as your lead holds the bottom.

BAIT

Probably more stripers are caught with sea worms (called sand worms, clam worms, or *Nereis virens*—depending on where you live) than with any other bait. Local regulations will control your digging, but if you are fishing commercially for stripers (especially if there are a lot of crabs to steal your bait), you'd better dig your own, because they can get awfully expensive if you have to buy them. Don't put the hook through the worm once and let it trail. That technique, with the worm hooked through the head, is for trolling. Put the hook through near the head, loop the worm, pierce it again, loop, and pierce it again.

A sea worm catches more stripers than any other bait.

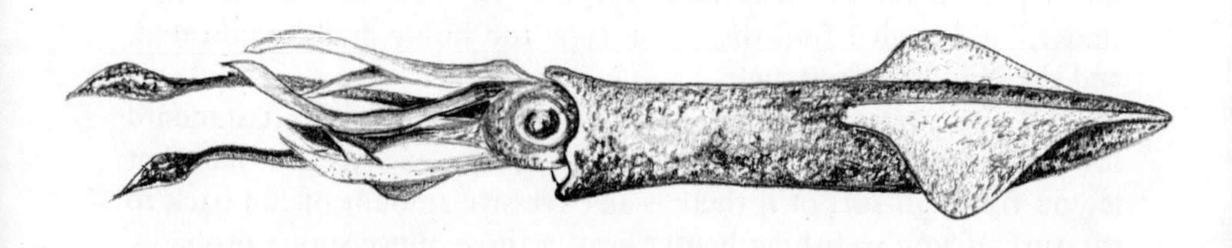

Squid, popular striper bait. Large squid should be chunked into two-inch pieces; small squid should be left whole.

Sea clams grow almost everywhere along the shore in loose, porous, sandy bottom. They are among the most economical bottom baits for stripers and flounders.

A more durable bait is squid, either a whole squid if you're using the little "summer" squid, or a two-inch chunk of the larger, "bone" squid. Fresh squid may stay on your hook longer than frozen, but the stripers don't seem to care whether your squid is fresh or frozen. Hook the bait through the squid and back again, so it hugs the hook.

After squid, use any of the "white" shellfish. My own preferences, in order, are: fresh razor clams—*Ensis directus* (frozen razors soften too quickly), sea clams (surf clams, hen clams, bar clams, skimmers, or *Spisula solidissima*—again depending on where you live), and quahogs (round or hard clams—*Mercenaria mercenaria*). While you may use any of the fishes for bait (except for sand eels, which are sometimes good in specialized places), most fish, particularly mackerel, will draw dogfish and ruin your fishing. The sole exception perhaps is herring (alewives—*Pomolobus pseudoharengus*). They are used whole and alive, or chunked up (the latter type is used only in the spring). You can also use live eels, but only under specialized conditions too complicated to go into here. Get some old-timer to teach you.

ARTIFICIAL LURES

The rules for sportfishermen and meat fishermen have been pretty much the same, so far, but now they differ. While the sportfisherman might be able to afford every artificial lure in the showcase, the commercial man must draw the line between delighting the tackle salesman and depriving himself of essential working tools. Artificials are all exaggerations of natural bait—bigger, brighter, noisier—so your choice should be keyed to the natural bait in your area at the time you are fishing. One rule holds: Use dark plugs of whatever color for dark days, bright plugs for bright days. While the surface or popping plugs are more fun to fish, the commercial man must remember that stripers will often hit a swimming or underwater plug, whereas they will run away from a noisy, surface-breaking plug. If you find a plug that works, carry a spare or two. Tackle *does* break down, and you can't afford to be where the fish are hitting and lose your only effective plug. You'll want to carry some "tin," some metal lures of two or three different kinds, particularly in the fall, to reach that school of fish that is breaking just beyond the reach of your best cast with a plug.

FINDING FISH

If you are new at the game, you'll probably have to go where the gang goes, until you learn your own water and your own stretch of beach. Pay attention to whether or not you catch fish, where you catch them, how many you catch, what the weather conditions are, what lures or bait you use, what time of tide and day it is. Write it all down in your little black book until it becomes second nature for you to catalog it in your mind.

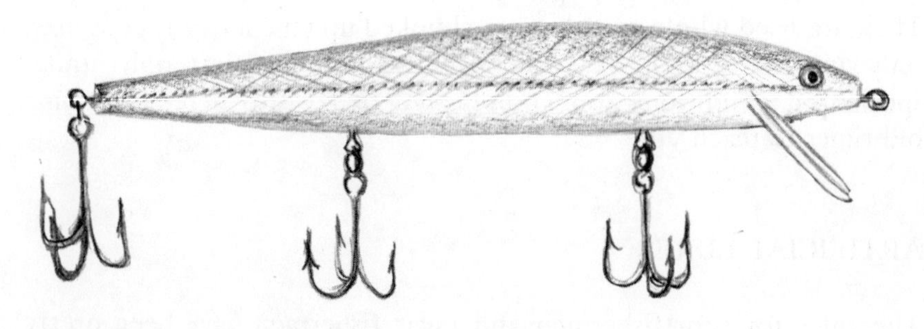

Light, spinning, diving plug.

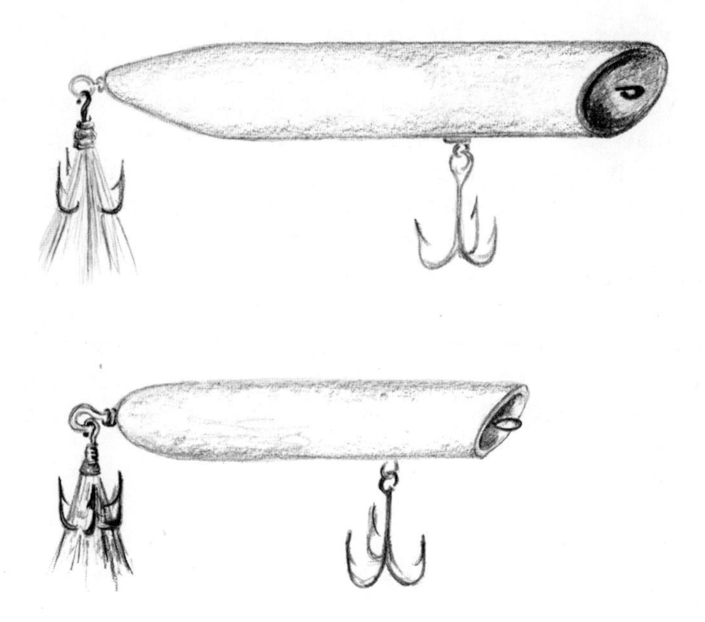

Surface or "popping" plugs are fun to fish. Balance them to your gear.

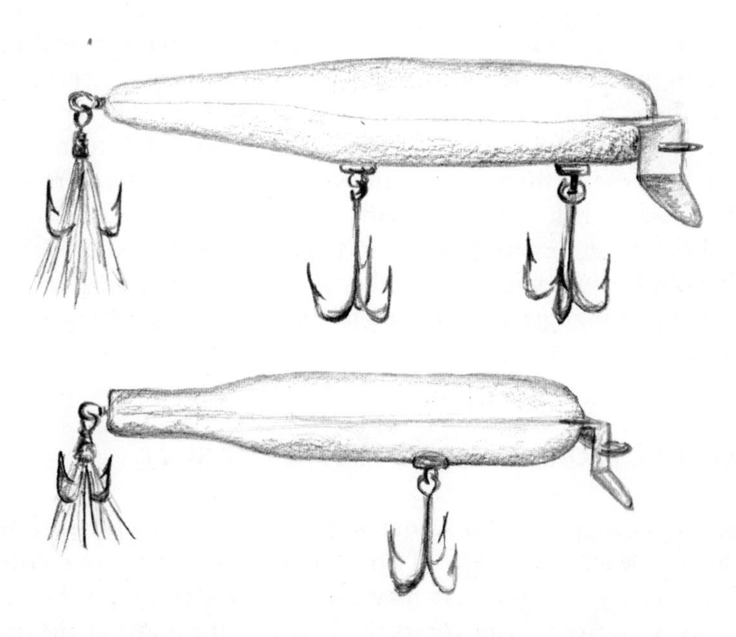

Diving plugs or "swimmers" will work well when the stripers are "down."

MAKING A LIVING ALONGSHORE

If you're searching for your own spots, remember that the news in the paper or on the radio is usually at least a day old. Even the gossip you pick up in the local tackle shop is about yesterday's fish, which may have moved before you can get there. However, you should study both the printed news and the local gossip for clues. If you go to the beach looking for likely spots, go at low tide. If possible, look down from a sand dune or other elevation with a pair of polarizing sunglasses. In general, stripers prefer a beach that is steep-to rather than flat. They like an inlet, or breakwater, or even a single big rock that causes an eddy, or a piece of slack water in which bait can hide or rest. If you fish a tidal estuary, look for a kink in the channel, or a sandbar over which the tide runs into a deep hole. Try to think like a baitfish, "Where can I go to hide and rest?" The stripers will find you.

WEATHER AND TIME OF DAY

Here again, the working fisherman differs from the sportfisherman. If, as many fishermen think, there is a slight edge at sunrise and sunset, then take advantage of it; you can catch up on your sleep the following winter. If an area seems to pay off during a certain tide, fish it then, but start earlier and stay later. I have never felt that stormy weather makes any difference in the way fish bite in the surf, but if it doesn't, then neither does fine weather. If you're hungry enough, you'll fish both. The only thing is, if the surf is so high you can't operate successfully, you can't make your plug work right, you can't hold bottom with your sinker if you're bait fishing, and *if the fish aren't biting,* get out of the surf and fish inside waters if any are available.

TWO RULES FOR BEACHING FISH IN THE SURF

One: don't backpedal. Work your way into or close to the surf before trying to beach your fish. Two: don't hurry; tire your fish out before you try to land it. It may take you a little longer, but a half-drowned fish, tired before it hits the beach, will pay off at the market better than one that was hurried and broke off in the surf.

TAKE CARE OF YOUR FISH

If you're planning to sell your fish, keep them cool. If you're in a skiff, lay them on their backs with their bellies up so they won't turn red. Cover them with a wet burlap bag if you have one. Bury them in wet sand if you're on the beach. (I guess I don't need to suggest that you mark the spot clearly so you won't lose it.) Wash your fish before you take them to market; very few fish buyers like to buy sand. Do not gut stripers before selling; most buyers prefer stripers in the round.

ETC.

Remember that there are other fishes in the surf that, while not as glamorous as stripers, often pay off better.

Flounders: mostly spring and fall; use smaller hooks, because they have smaller mouths; use sea worms or shellfish for bait (sea worms or even garden worms will consistently catch more fish, but shellfish also will consistently catch bigger flounders); do not use squid or other fish bait.

Fluke or plaice fish: mostly summer fishing; they have teeth, so use wire leaders; squid, fish chunks, or even live minnows are more productive bait than worms or shellfish.

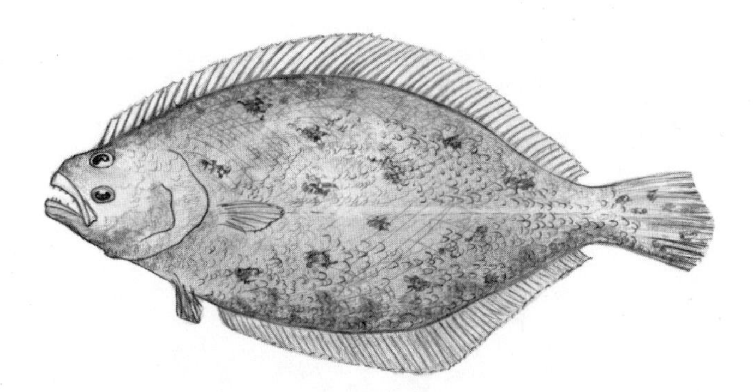

Fluke (or plaice fish or summer flounder) may grow as big as 20 pounds, but they are more commonly only five to 12 pounds. Use wire leader, and some sort of "fish" bait—live minnows, if you can get them.

Codfish: use striper-size hooks; use almost any bait, depending on the area, even artificial lures at times.

Bluefish: mostly summer and fall, and not every year; wire or cable leaders are a must; use almost any bait except, perhaps, squid; use artificials if you can find the fish top-of-water.

Weakfish or squeteague: missing from this coast for 40 years, they seem to be on the way back; again, a wire leader is safest; use almost any bait a striper will take, natural or artificial; move the bait gently through the surf, *gently*: they're called weakfish because their mouths are so tender.

Squidding with tin in the surf for mackerel or pollock can be fun, but it's seldom very profitable. However, keep an open mind; there are times when big, "number one" mackerel pay off handsomely (particularly if there are bluefish around). If sold locally, they can be very worthwhile, but keep them cold and wet or they'll spoil.

Almost all these fishes—particularly mackerel, bluefish, and squeteague—must be cared for lovingly or they'll spoil on the beach before you can get them to market.

2 FLOUNDER DRAGGING

Flounder dragging in inside waters is a sometime thing, and it may even be outlawed by the time this is printed, but this chapter is included because it was a way of life and it may still be legal in some areas. At this time, it is permitted in certain estuarine waters in Massachusetts, from November 1 to May 1, provided you apply first to the Massachusetts Department of Natural Resources, Division of Marine Fisheries.

Besides a 16-foot or longer skiff with an outboard, the equipment used these days is a miniature otter trawl or fish drag, varying in width from 12 to 35 feet. It is not an exact duplicate of the offshore trawls, because conditions are different. Tide run and chafing on the bottom are not all that important, but the estuarine dragger faces a problem usually unknown to his offshore counterpart—an overabundance of marine weeds, eelgrass, sea lettuce, and, worst of all, mermaid's hair. Furthermore, the outboarder does not have the rig to power-haul his load aboard, splitting his net as he goes. If his load is more than he can haul in, singlehanded, over the side, he has only one recourse, and that is to tow his load ashore and unburden himself of unwanted weeds and trash.

The present flounder dragging law on Cape Cod admittedly is wrong, and it has been wrong ever since it was put through the legislature by lobstermen trying to protect their pots from inshore draggers. Flounder dragging should be allowed in inshore waters by October 1, and it should be outlawed not later than February 1, or

March 1 at the very latest. After March, even in a very cold year, the flounders have run into clean, sandy-bottom, shallow water to spawn, and they should not be disturbed. Also, in the spring the flounders are thin and watery and a very poor imitation of the fat fall flounders. A bushel of flounders in the fall weighs upwards of 70 pounds; the same bushel in the spring weighs scarcely 60 pounds.

However, quite significantly, the salt ponds where flounder dragging has been outlawed completely no longer produce flounders, not even for the handliners in the fall. The adage, "If you don't cultivate a garden, you can't grow any vegetables," holds true in this case more significantly than in almost any other type of fishing. The doors of a drag plow up the soft mud bottom, oxygenate it, sweeten it, and make it more productive of both fish and shellfish.

BUILDING A FLOUNDER DRAG

There can be a lot of variations in making a flounder drag, but the dimensions should be proportionate. The drag described here is 12 feet wide and 15 feet deep, and made of four-inch nylon mesh. It is made to be towed behind an outboard and hauled by hand. The wings are additional, and they may be extended as far as you like.

If you "frame" the drag with quarter-inch nylon rope, you will find it more durable and easier to tie. For the bottom of the drag, stretch out a kind of truncated rope triangle over several clotheslines or extend it to each of four convenient trees. It should be 12 feet across the front or forward end, and 15 feet deep, and about 4½ feet wide at the tail or back. This will be the size of the drag; top and bottom are the same. The sides will be two feet (or eight meshes) high along the whole length.

You will need two pieces of netting, one for the top and one for the bottom. Each should be 45 meshes wide and 48 meshes deep. Each of the two side pieces should be eight meshes wide (as noted above) and 56 meshes long, or longer, depending on how long you make the wings. (As far as I'm concerned, long drag wings are a carryover from offshore drags. They make hauling back more complicated but they don't help increase the catch.)

Start off with the truncated triangle of quarter-inch nylon rope and a corner of the bottom netting. Sew or whip the netting evenly

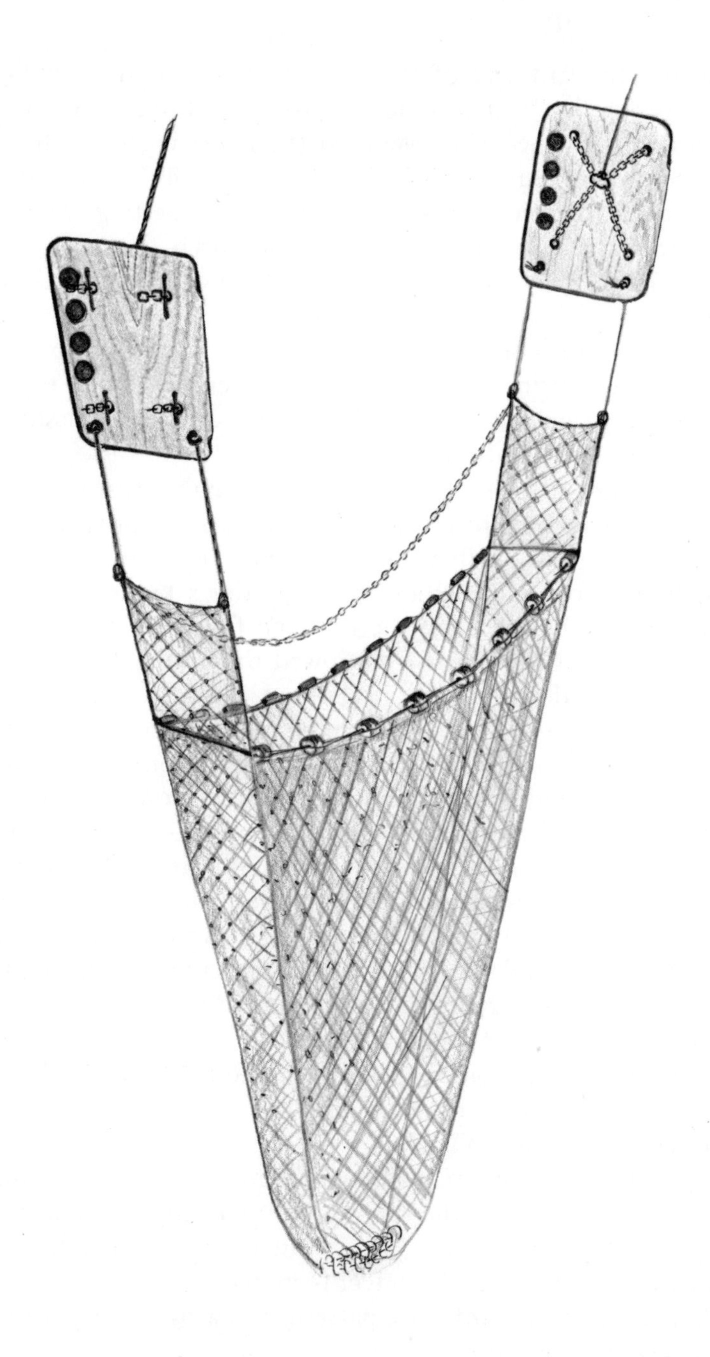

A complete flounder drag.

across the forward end of the bottom. With four-inch mesh (two inches to the "leg"), this comes out to about three to 3½ inches to the mesh. Stitch the twine evenly at the knots in the mesh, and try not to make any bunches, which will collect gobs of bottom weeds while you're dragging. I like to whip on the netting with a clove hitch, locking it with an extra half hitch in case of later chafing. I use the same weight twine for sewing as is used for the netting.

Start at the port side forward corner of the bottom section and whip the netting to the standing quarter-inch nylon rope every three or 3½ inches until you've whipped on 45 meshes. This should cover your 12-foot width.

Now kick out or stretch out your netting toward the stern end of your strengthening rope. Again, sew or whip your twine onto the side ropes, proceeding first along the port side. But here you have to taper. Sew on two meshes at the netting's knots. Now cut out two legs of the next square, the first and fourth as you work down the netting. You should have narrowed the netting by half a mesh. Work this way down the entire 15 feet. Whip on two, cut out the third, and whip the rest close. Whip on two and narrow one, keeping the twine spread evenly. By the time you have gone back 48 meshes, your netting should be narrowed a half mesh for every third mesh.

Go back to the starboard side. Again working toward the stern end, whip on two and cut one (the first and fourth leg of that mesh). When you have finished whipping the bottom onto both sides, the bottom should have 45 meshes across the forward end and 28 meshes across the stern end. If you are a few meshes short, don't let it bother you, as long as the netting is stretched evenly fore and aft. The flounders won't know the difference.

Sew the top section exactly as you did the bottom section. The most important requirement is that the net be smooth, *smooth,* SMOOTH, from front to back.

Now take a piece of netting eight meshes wide and whip it to the bottom and then to the top, smoothly, from *stern* to *bow.* When you get to the forward end, let it run by to the length you have decided on for the wings. Two feet is enough as far as I'm concerned, but if you follow the offshore pattern, the wings can run forward as much as six feet or more.

In order to have strengthening ropes for the wings (this is where the strain will be), you'll have to attach more of the quarter-inch nylon rope to the basic frame ropes. You can tie in this rope if you want, but you'll be sorry, because every knot is one more mesh-catcher. It is much better to splice it in. If you are going to have two-foot wings, you will need six feet of rope, plus enough for two splices, plus enough for "ears" or loops on both the top and bottom corners of the wings. These ears are used to attach the ropes leading from the "doors" or "boards," which will spread your drag when it is towed. In all, you will need about 12 feet of rope for each two-foot wing.

If you are using nylon or one of the new "poly" ropes, you will have learned by this time that the stuff "explodes" when you cut it. It may be lubberly, but I usually tape each of the three strands to make splicing easier. Masking tape will do if it's only temporary, or you can use black plastic electricians' tape.

Start at either forward corner and lay out your wing rope parallel to the length of the drag, allowing about a foot for your splice. Here, again, it's easier to tape the two ropes together, at least temporarily. Unlay the wing rope to the tape, or until the middle strand is on top. Tuck that strand under the middle strand of the lengthwise rope. Tuck the left-hand strand under the left-hand strand of the lengthwise rope. Twist the splice over and tuck the right-hand strand of the wing rope over the right-hand strand of the lengthwise rope and up under. As in any splicing, each strand should be led out between two different strands and opposite each other. Pull all three strands in, so the ropes lie smoothly, then tuck under one and over one in rotation, pulling each tuck tight and making sure it is smooth. After three tucks on each strand, you should have enough, although the "poly" ropes are so smooth that you may need more. Then lead the loose wing-rope ends toward the stern end of the drag, twist them together, and tuck the three of them between any two strands of the framing rope to lock them. I find this safer than cutting off the ends.

Lead the wing rope forward three feet (for a two-foot wing) and loop it back on itself. About a foot from the doubling, whip the two ropes together securely. (You can try eye splicing in the middle of the rope if you want to, but it's tedious and bulky.) Then lead the rope down two feet, make another "ear" or loop, and come back to secure it into the forward corner of the bottom section.

To these squares of strengthening ropes, whip the wing netting as you did on the sides of the drag, again without tapering. (If you have made a drag deeper than two feet, you might want to taper the wings to fit the height of the boards, but be sure to taper smoothly.)

You will want a head rope with some flotation capability, and a foot rope with some weight. On a 12-foot-wide drag, I used 10 net corks for the head rope. Thread another length of quarter-inch nylon rope through the net corks, splice the end of the nylon into the wing-strengthening rope, then whip this latter rope along the leading edge of the head rope, spacing the corks evenly. I use the same clove hitch with a locking half hitch—making sure each time to catch the corresponding mesh of the net—and splice in the far end.

The foot rope is done the same way, but I used 18 one-inch roller leads on the foot rope, the same rollers you would use on a scallop drag, threading them on another length of quarter-inch nylon rope. It is spliced and whipped along the leading bottom edge, as was done with the head rope.

There is a great deal of argument among fishermen about weighting the foot rope. If you are determined to do it, you can whip on a piece of light chain 3/16 inch or more. The first drag I started with had almost twice as much chain as the width of the drag. The chain was looped back about six inches for every foot of width. I could never understand how this chain could help catch any more flounders, because the loops dragged *after* the leading bottom edge and simply stirred up more bottom vegetation.

A "kicker" chain makes more sense, particularly if you are dragging in bottom with a lot of hangups. This is simply a chain that is about a foot shorter than the leading edge of the bottom of the drag. It is secured on each side to the leading bottom corner of the wing. The theory is that it is dragged along the bottom just slightly ahead of the drag and "kicks out" the flounders. Whether that is true or not, it will save your drag from being torn up on hangups —lost anchors, old cedar stumps, and similar trash.

All that remains to be done to finish the drag itself is the tail end. If you are going to haul with power and have to put a strap (a double loop of rope) around your load, then you'll need a "cod end," an extra bag the same width as the drag itself and maybe six feet long, but of heavier twine. It should preferably have some sort of chafing gear on the bottom to save wear and tear. But if you are hauling by hand over the side of a skiff, I see no reason for the cod end. At the back end, simply sew a two-inch ring on every mesh. Through these rings, in order, reeve a length of the same quarter-inch nylon that was used to frame the drag. (The rope should be a fathom or more longer than you can spread the opened back end of the drag.) Splice on another ring at each end of this rope. The pucker-string knot is simply a chain of half knots that can be released, no matter what the pressure, merely by pulling first on one end and then on the other. Tuck a loop of the right-hand end through as though you were going to tie a square knot, and pull both ends tight. Then tuck a loop of the left-hand rope through *that* loop and pull the right-hand rope tight around that, alternating ropes as you go. A half-dozen loops should do. The drag is now finished and you are ready to make the doors or boards.

The doors or boards for a small flounder drag spread the drag out and down as it is towed along the bottom. These boards are temperamental, and no two act exactly the same, no matter how carefully they are constructed. Their big brothers on the offshore draggers, of course, are built of three-inch oak and sheathed with iron. My own boards are made of ¾-inch exterior plywood. They each measure 20 inches high and 36 inches long and are bound on the bottom, forward, and aft edges with ½-inch half-round iron. The boards are drilled about every six inches for two-inch #8 flat-head

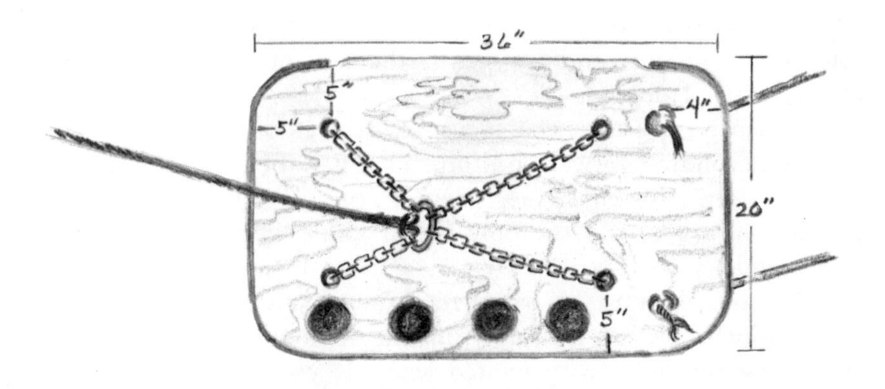

Front view of door of flounder drag.

screws, which are countersunk so the surface is smooth. All four corners are rounded off on a six-inch arc.

One-inch holes are drilled on the face of each board, five inches back from the leading edge and five inches in from top and bottom. Two more one-inch holes are drilled eight inches forward of the after edge. One is five inches from the top and the other is five inches from the bottom. It is through these four holes that the leading chains are adjusted and secured. Also, four inches forward of the after edge, and four inches in from top and bottom, are two ⅝-inch holes. Through these are rove the ropes that lead to the forward ears or loops at the ends of the wings.

The four leading chains—one through each of the four one-inch holes—can be of 3/16-inch chain, and the inside ends of all four should be secured to some sort of ring, to which the drag warp is also tied or spliced. I had the local blacksmith weld this ring in place. The other ends of the chains are led from the inside of the boards, out through the holes, and they are secured on the outside. Here again, we went to the local blacksmith and had him make eight pins of ¼-inch round iron, each three inches long and flattened on both ends. Through these flat flanges we drilled holes big enough to take one-inch #8 screws. Each pin could be slipped through a link of the chain and then secured to the board on the outside, thus making the easiest, smoothest, and most positive adjustment for the length of the chains. The forward chains should be about 18 inches long, the after chains about two feet. By shifting the pins, the chains can be shortened or lengthened, allowing the forward end of the boards to pay out farther to the side or be pulled more nearly straight ahead, and the tops of the boards can be tipped farther out or pulled more nearly to a vertical plane.

Presently, my own chains are set to pull out and down. They are set as follows: forward bottom, 11 links; forward top, 15 links; after bottom, 20 links; after top, 21 links. These numbers are variable, however, and the only way I know to get the set right is to cut and try.

The boards must be ballasted. I solved this problem by boring four 2½-inch holes an inch up from the bottom edge of each board and filling the holes with melted lead. (If you prefer, you can cut in a strip of lead as thick as the plywood, two inches high and 20 inches long, and secure it by boring through the iron binding strip, through the lead, and up into the plywood a couple of inches in half a dozen places.)

The ropes from the boards leading back to the drag can be half-inch nylon or "poly." Three feet is a handy length, although some draggers prefer as much as six feet to give a wider spread to the boards. It is debatable how many flounders kicked out by the boards go between the boards and the net. This wider spread is the logical reason for longer wings, though I never found that it made much difference in the catch in inside, estuary dragging, and the wider spread certainly collects more marine grasses. The ropes should be tied with a common or round knot both inside and outside the boards, with the long end of the rope leading inside. Both ends of the ropes should be whipped to prevent fraying, particularly the after ends. The after ends should be led equally through the forward ears on the wings and tied with a sheet bend with an extra turn tucked through to prevent jamming. The head rope may be shortened slightly by shortening the top ropes from boards to drag, if you feel the flounders are lively and apt to go over the drag.

The length of the two drag warps, of course, will depend on the depth of the water. Since I seldom drag for flounder in water deeper than four fathoms, I usually plan on 12 or 13 fathoms of warp on each side, or a ratio of slightly more than three to one. Again, this is variable, and you'll have to work out your best length; it's better to have too much rope than not enough. I like to use half-inch manila—since it is somewhat easier on my hands than the synthetics—and secure each tow rope to a cleat inside the rail, well down and out of the way, so the net will not catch on it on the way out.

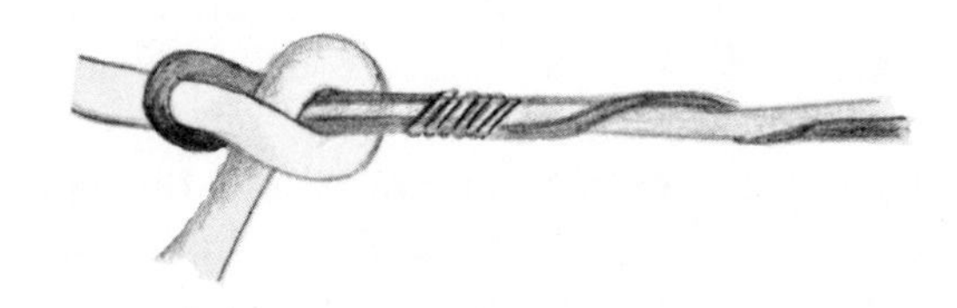

Sheet bend used to connect wings and door rope of flounder drag.

Before you start, see that your gear is in order. First your drag warps are coiled down together. Then the boards are set up with starboard board well forward, the port one aft. Between them, the drag itself is faked down, foot rope and head rope abreast of each other, the pucker string last and on top. You are ready to set out.

With the skiff going ahead, cut in a wide circle slightly to starboard, chuck over the pucker string and the heavy rings: they will pull the net along, though you may have to help at first. Then over the side with foot rope and head rope. Be sure everything is clear to this point. Then flip over the starboard board, bottom edge *out and down* so the board dives away, then flip over the port board the same way, only with the bottom edge *in and down.* Let the ropes run while you straighten out the boat. If you haven't fouled anything, the drag warps should spread evenly, making a nice, long triangle.

As in all dragging, conditions will govern your speed. If the water is relatively warm, you will have to go faster. If, on the other hand, the water is very cold and the flounders have bedded down deep in the mud, you may have to slow down; you may have to set the boards so they'll dig a little deeper. Using the gear described, with a 16-foot skiff and an 18-horsepower outboard, you'll need most of your power.

Don't make the first tow too long. Try towing for 15 minutes, and then haul back the gear to find out how the drag is tending bottom and how much trash you are picking up. Throw the motor out of gear and haul back both ropes at once, dropping them together in the bottom of the skiff. When the boards break the surface, lift them over the side, one at a time, the starboard board away forward, the port one across the stern. Then haul in the foot rope, draping it evenly along the inside of the rail. Bring in the head rope quickly. Once the foot rope is in, everything you've scooped up in the net is yours. Shake the net as you haul it aboard to shake down the grasses and fish, faking it down evenly after it is shaken out between your knees and the rail. When you get to the end, you may want to strap the net together to keep flounders and trash from rolling back inside the net. I usually use a heavy piece of rope, ¾ inch or so, spliced into a circle about two feet across. Tuck one loop through the other around the net and jam it down. Then bring in the rest of the net, rolling it into the boat. Pull the pucker-string end clear of your faked-down net so it won't interfere with setting again, then shake what you have caught onto the bottom of the skiff.

Re-tie the pucker string—carefully. It's pretty exasperating to make a whole tow, haul back, and find everything has gone out through the open bottom end. Take off the strap and put it out of the way. You're ready to tow again, circling to starboard as the drag goes over the side, but straighten out the boat before you have doubled back over the net. You'll soon learn to haul back into the wind when you take the drag, or else you'll be apt to drift over the drag and get the twine tangled in the outboard propeller.

Once you are towing again, throw or fork over all the trash until the skiff is reasonably clear. Cull the flounders two ways: put those that weigh about a pound in one basket, those more than a pound in another. Don't keep anything under 12 inches long; they will add little to the total weight of the catch and will knock your price down. Furthermore, if you take all the babies this year, what are you going to catch next year?

As in all fishing, wash your catch well—it will improve the price. If you are packing the flounders in flounder boxes, pack them belly up, the white side up. As with stripers, it keeps them from turning red. You may want to save a few "give-away" flounders, under 12 inches. A dozen or so won't make much difference in next year's crop, and they'll make a surprising number of friends.

3 EEL POTTING, SPEARING, AND SKINNING

Very little is understood about the life history of the eel, *Anguilla rostrata.* To quote *Fishes of the Gulf of Maine* (Fishery Bulletin number 74 (1953) of the Fish and Wildlife Service, Henry B. Bigelow and William C. Schroeder, U. S. Government Printing Office): "Eels grow slowly. . . . Eels tolerate a wide range of temperature . . . and salinity. . . . Eels seek muddy bottom and still water. . . . But this is not always so whether in salt water or fresh. . . . Eels are chiefly nocturnal in habit. . . . But eels, large and small, are so often seen swimming about, and so often bite by day that this cannot be laid down as a general rule. . . . No animal food, living or dead is refused, and the diet of the eels in any locality depends less on choice than on what is available. . . ."

Eels taken on hook and line are generally a nuisance, although there are fishermen who deliberately fish this way for eels. If you are so unfortunate as to hook an eel while bottom fishing for flounders or stripers, either cut the hook off and let the eel go, or give the eel a sharp blow just forward of the leading edge of its dorsal fin, then grab the eel with the first and third fingers under and the second finger on top, a little back of its head, to hold it while you are unhooking it.

"The greater part of the [eel] catch," to quote Bigelow and

Schroeder again, "is made in nets and eelpots; and some are speared, mostly in late autumn and winter, often through the ice."

The use of eel fykes pretty much went out on Cape Cod with the drainage ditching of the salt meadows by the mosquito-control projects: the eels spread out in a hundred different directions and were impossible to trap. The old fykes were made of spruce laths. They were set in a creek leading to a freshwater spring at the head of the meadow. Wings were set out across the marsh on both sides, a couple of feet high and several feet wide, pitched toward the center rather than at right angles to the creek. In some cases the laths were driven into the marsh grass, half an inch or so apart, and cleated at top and bottom, or the cleats were driven into the marsh, spaced a lath-length apart, and the laths were nailed to them horizontally, again about a half-inch apart. These wings should be set securely, because the high-course tides in the fall bring mats of thatch and eelgrass sweeping in over the meadows to clog the wings and carry away all but the rugged ones.

In the creek itself, if it was not too wide, a square funnel was built with laths spaced about a lath's thickness apart, nailed to two-by-fours or even heavier timbers. The open, downstream end stretched across the entire width of the creek; the smaller, upstream end was no more than an inch on a side. The downstream end was flush with the bottom of the creek and properly footed with gravel or broken stone, partly to prevent erosion and partly to prevent the eels from finding their way under the funnel.

The bottom of the upstream end of the funnel was placed a

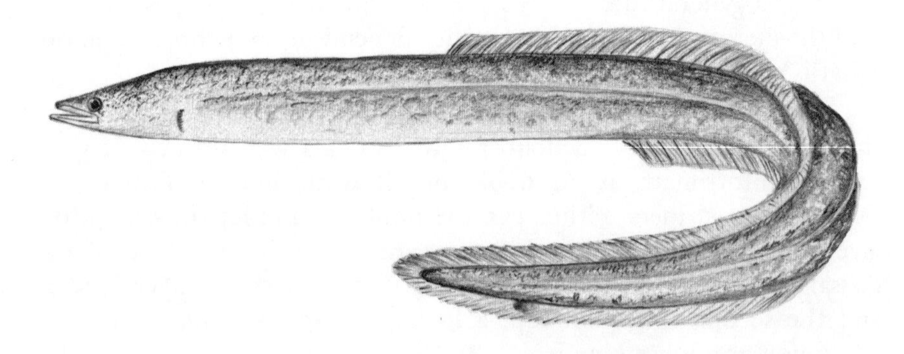

Little is known about the life history of the eel, *Anguilla rostrata.*

foot or more above the average low-tide level, so the eels could not find their way out when the tide was down.

In back of the funnel, and nailed to it, was the car, into which the eels could find their way through the funnel, and from which they could be bailed at low tide. It, too, was made of closely spaced laths, stoutly fastened. It rested on the bottom and was as wide and as high as necessary to fit the funnel—three or four feet or more, and as much as six feet long. At the top was a latched door through which the eels could be dip-netted.

Since this is an old-time rig (the last one I saw in operation was 40 years ago), I don't know whether or not it will work today. I don't see why not. You can expect your major catches on the first frosty nights in the fall during October.

I set an eel pot behind the funnel of an old-time fyke some years ago, and I filled the pot that I used in place of the broken-down car on several nights. I had some difficulty getting the eels across the marsh. A burlap bag, I found, was no hindrance to the eels. They could push their tails through, and half of them were lost before I could reach high land. Putting one burlap bag inside another was somewhat better, but eels still escaped. I suppose a stout, fine-mesh bag would be the proper transport, but remember that if you keep eels too warm or cut off their air—as in a tight plastic bag—they will suffocate.

EEL POTS OR TRAPS

While there have been some changes in design over the years (seldom for the better), eel pots are basically the same. Usually cylindrical in shape, eight to ten inches in diameter and 30 to 36 inches long, they have either two or three heads or funnels.

Let's start with a basic pot, which may be altered by preference. Nine inches in diameter and 30 inches long, it is made of ⅓-inch mesh, galvanized hardware cloth or "cellar-window wire." It is framed on three nine-inch hoops made of ¼-inch galvanized rod. (These hoops were hard to come by in the old days, and they are not always available now, so we used to resort to several turns of galvanized telephone wire—with enough stiffness to give strength to the covering of the pot. One of these hoops was secured to each end of

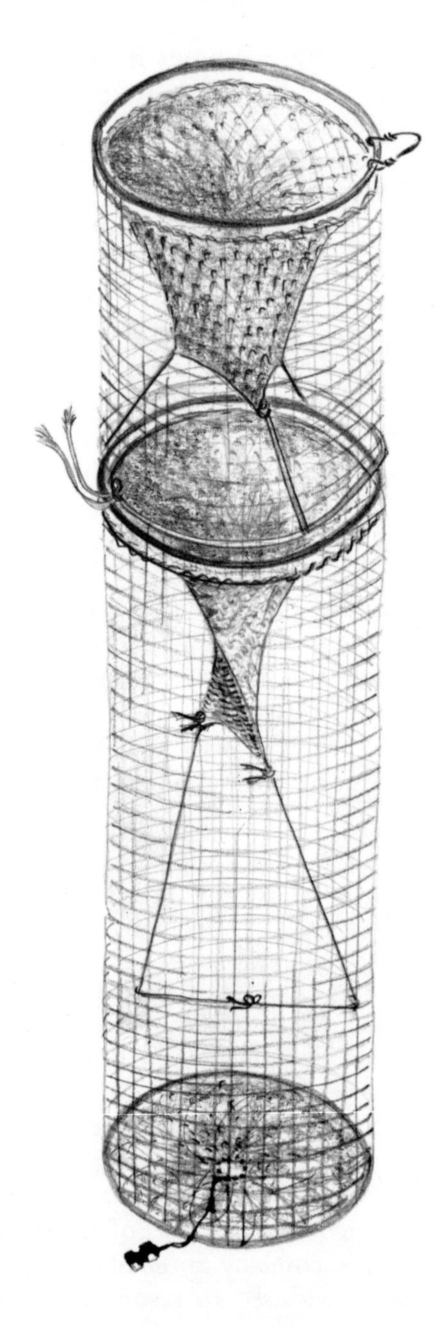

The basic design of eel pots has changed little in the past 40 years (and seldom for the better).

the pot in a half-dozen places with soft copper wire. In later years, fearing an electrolytic reaction, we used fine galvanized wire in place of the copper, but I couldn't see that the pots lasted any longer, and the "galvy" was harder to twist tight.)

The third hoop is wired in place a foot back from the leading end. The two lengthwise edges of the hardware cloth, one edge lapping or running by the other by a couple of inches, are secured in three or four places with wide-burred copper rivets. Without these, the edges can be wired together, but the pot won't be as rigid.

The leading head or funnel is sewn in just aft of the forward framing ring. Since the heads we used were made of treated cotton, we used six-pound codline for sewing. The modern nylon heads may be threaded on the forward framing ring, or sewn on with fairly heavy nylon. It is best to catch the head temporarily in three equally spaced places before the threading or sewing is done. The meshes should be spaced as evenly as possible around the pot.

A heavy crochet hook with a built-up handle is good for hooking out each mesh. Push it through the meshes of the wire, hook a mesh of the head, and pull it out through the wire. Thread the codline or heavy nylon through the head mesh. Repeat this process all around the pot, being careful to space the head meshes as evenly as possible with the wire meshes. Try to skip no more than one wire mesh at a time, because an eel can work its tail through any small hole. After the codline or nylon is pulled tight and the head is even, tie it off with several knots. The knots will loosen if they're not tied very carefully.

Divide the inside edge of the head into three equal parts, and, gathering two meshes together for strength, tie at each of the three corners a length of codline, which must be led out through the wire, just forward of the second ring. Pull these tight, stretching them with your hand if necessary, and tie them all together. You will find that the inside end of the forward head has opened up some four or five inches on a side.

The middle funnel should be longer than the first one. It should be sewn in just aft of the second ring (which should be secured to the pot no more than a foot aft of the forward head) the same way as the first funnel. But at the after end of the middle funnel, tie two instead of three leading strings opposite each other. These should be stretched back and out through the pot near the after end, and then stretched tight and tied securely. This will give you an opening or lip through which you can push your fingers or an eel can make its way, but it will close after an eel has gone through.

The third head should be like the first, and it is sewn into the pot the same way, but around the inside edge is threaded another length of line—not necessarily at every mesh, but at least at every other mesh. The old-timers taught me that you could never pull this pucker string quite tight, so use a wooden spool and pull the pucker string around it as tightly as possible. Rather than use a machine-made spool, we took a two-inch-thick cedar sapling and cut it into two-inch-long pieces, and then cut a deep groove around the middle of each strip. It's handier if the spools are tied to the pot with a foot or so of line. It keeps them from being lost overboard.

At first we used to tie the buoy line directly to the pot, but pushing six-thread rope through ⅓-inch mesh meant spreading the wire, so we made a small loop of telephone wire on the forward end of each pot. There used to be an argument that the buoy line should be tied to the after end of the pot, the reasoning being that if the parlor or center head let go when you were hauling, you would still have a chance to save some of your eels. However, if there was any appreciable tide-run, the buoy line would tend to pull the pot in the direction of the tide, and it was thought that eels feed pretty much heading into the tide. I don't know that it really makes much difference; it's one of those things you have to work out for yourself. I finally settled on tying the buoy to the forward end. The buoy line, of course, should be long enough to reach to the top of the water at high tide, plus a little bit of slack. Better too long than too short. Since the pot is round, there is no need for a bottom float to keep the rope from hanging up, as in lobstering. For a buoy, we finally settled on a nine-inch piece of two-by-four, having messed around with net corks on the one hand and larger buoys on the other. You don't want too large a buoy, because the tide will tend to drag it, particularly if it gets fouled with drifting grass. On the other hand, you want a marker large enough so that you and the passing boatmen can see it from a distance.

As for variations, a smaller pot in most cases will catch as many eels as a larger one, is easier to stow and handle, and will stand up longer. Quarter-inch cellar window wire is all right if you're fishing for shoestrings for bait, but it clogs easily and thereby rolls more easily with the tide. Half-inch mesh allows cleaner fishing and will release some of the smallest eels, which you probably don't want anyway, but the wire tends to break down more quickly, particularly if you're catching a lot of eels and make a practice of resting the pot crosswise on the gunwale while you tend it.

Since eel-pot heads made of twine are expensive and sometimes hard to come by, some of the local boys substituted a length of a woman's nylon stocking in place of both the forward, open head and the center funnel. It works well, except that the pieces of stocking are hard to secure to the wire, since they usually are whipped over and over with heavy thread and a sail needle.

Another substitution was a funnel made of cellar window wire to fit the leading edge of the pot and extend about six inches into the pot, tapering in that distance to a two-inch, circular opening. From this, a two-inch tube of the same material led almost the full length of the pot, stopping only a couple of inches short of the back. The tube was securely centered. The after head was replaced by a hinged door, made of the same material, overlapping the pot edges and securely closed in two or three places. If there is a weak place, no matter what kind of pot, the eels will find it.

In the old days, when weirs were common along this shore, all eel pots were dipped in hot tar, as was the twine for the weirs. In cold weather the tar hardened, but it was softened by being kneaded in lard. I have fished the wire type of head as well as the softer cotton heads (or now, nylon), and it seemed to me that the eels preferred the softer ones. The softer heads work better. (Since writing the above I have talked with a working eeler. He agrees with me, and to compensate for the harsh wire, he lines the tube with a length of his wife's worn-out body stocking, leaving the after end loose and flapping.)

Going even further, when wooden nail kegs were commonplace (we used to be able to buy all we wanted second-hand from the local lumber company for a dime apiece), I knocked out the bottom head of the keg and replaced it with a double thickness of white chicken-feed bag, locking this on with the bottom hoop. The keg was ballasted with a couple of bricks wired in place. The forward end was extended some eight inches with cellar window wire and headed with a funnel-type head. The keg pots caught a lot of eels, but they were heavy and hard to handle, and they tended to roll with the tide.

How many pots make a working string? That depends entirely on how hard you want to work, how big an area you have for fishing, and how much capital you have. I knew a successful old-timer who carefully spotted 17 pots and let it go at that. My partner and I ran 165 at one time. To my way of thinking, that was too many, particularly if you have to take time out to get your own bait.

BAIT

You'll have to experiment somewhat with bait. In most local areas, we found that chopped-up (two-inch strips) female horseshoe crabs made the best bait. However, in one area where there were several shellfish draggers tying up at night, washing down their decks, broken shellfish had a slight edge over horsefeet. If I were potting where fish draggers or line trawlers were tying up, I would first try any kind of fish available. While it is true that eels are scavengers and will probably eat anything up to and including a dead horse, you will find that in the long run eels prefer fresh, clean bait. And don't throw your used-up bait overboard where you are potting; dispose of it somewhere where the eels can't get at it. Male horsefeet will work, though not as well as the egg-bearing females. Chop them into chunks small enough so you can get them into and out of the after end of your pot without too much trouble. If you are short of bait, mashed-up conks or moonsnails are better than nothing, or mussels, or scallop guts. Of course, you'll be putting your bait in the parlor, the end aft of the middle head.

A horseshoe crab is good eel-pot bait.

In every area you'll find that there are spots where eels feed and spots where you'll catch mostly migrating eels. Eels feed under docks or piers, if there's not too much tide run. Look around the edges of a channel rather than in the middle of it (which makes it handier to keep out of the way of outboarders and trolling fishermen). If there is heavy eelgrass, go at low tide when the grass tops are near the surface, and you will find that eels have paths or roads that they follow consistently. It's almost impossible to catch all of the eels in any given place, but you could thin them down so drastically that it would not be profitable to pot there. Come back in a week or so and you may find that the area has become productive again. They do not pot well in spring holes, where they bed down for the winter.

There is no reason you shouldn't start to pot eels by early or mid-May, particularly if you are looking forward to selling live eels to bassfishermen or bluefishermen. We usually had more profitable things to do, so we didn't start potting eels until after Labor Day. Eels will have largely stopped running by the beginning of November, although I have potted them almost to Thanksgiving Day.

Handling potted eels, especially if you are catching 300 or 400 pounds a day, can be a problem. When we first started, we moved them directly from the pots in salt water to cars, or holding crates, in fresh water—and figured on a 25 percent loss. Then we discovered that if you hold freshly potted eels in cars in salt water for a week or so, they will disgorge all the food they have not digested and the survival rate will climb wonderfully. If you handle them properly, your loss should be negligible.

But handling them properly during hot, August days in a small skiff takes some doing. We tried turning them loose in the bottom of a 16-foot skiff as we worked, bailing out the old water and slime and bailing in fresh water as often as possible. This procedure worked fairly well, but it was messy and time-consuming. Don't put too many eels in any one container, for a start. We finally settled on half barrels, filling them no more than half full of eels—50 pounds or so to a container. These half barrels were low enough so that we could stack on top of them any spare pots we might be moving, any baskets of bait, used or unused, and still carry six or eight half barrels in the same 16-foot skiff. As a matter of good practice, if the run of eels is heavy, it will pay you to stop part way through your string and unload every three hours or so.

When your eels have cleared themselves in your saltwater cars, move them to fresh water, where they'll be safer, culling them as you do. Separate them into: shoestrings (release these unless you have a market for them for bait); small; and over a pound. If you tip your half barrels slightly to let the eels back out, you may be amazed at how easy they are to handle. An eel that's been hurt or frightened will tighten up its belly muscles and you'll have to stun it to hold it, but an eel backing slowly out of your keg will glide gently into your hand, its belly still soft, and you'll have time to carry it to whichever container it belongs in.

When I first started holding eels in cars for the Christmas market, I took the old-timers' advice and built slatted cars big enough to hold a half ton without crowding. They were hard to handle, and I didn't get smart until a slat came off one of them, releasing half a ton of eels into the pond.

Then I built much smaller cars, each holding about a barrel of eels. I made a framework of spruce strapping, three feet by three feet by 18 inches high, and covered it with half-inch-mesh cellar window wire, well stapled. Outside I nailed (with the pieces at right angles to the flat, inside framework) another frame of one-by-three strapping. The cars were light, easy to handle, relatively inexpensive, and lasted as long as five years. The tops, which were removable, were simply secured with battens nailed securely, but not so securely that I couldn't pull out the nails when it was market time.

The pots should be tended once a day. The catch should be taken out and the old bait should be replaced with new bait. In hot weather we tried to do this as early in the day as possible, spending the afternoons getting bait for the next day. There are purists who contend that pots will fish better if they are taken ashore to dry during the day and then reset each evening. Truthfully, I can't say, because I never had time to try this method. It could very possibly work.

EEL SPEARING

Theoretically, eel spearing should never be permitted on Cape Cod before the first of December—or until the water is thoroughly, *thoroughly* chilled—but most shellfish constables, under whose jurisdiction (for some unknown reason) eeling comes, feel the business is not important enough to warrant further restrictions.

In the strict sense, geologically speaking, there are no freshwater springs on Cape Cod, because a spring by definition is an upwelling of fresh water from an underground source through rocks and/or gravel. What are called springs locally are actually upwellings of rainwater through mud or sand. For our purposes, they might as well be called springs, as they are, and the softened places in the bottom where the upwellings occur are called spring holes.

When the water chills enough, the local eels seek out these spring holes, since the water there is warmer than the surrounding salt water. They bed down, moving only occasionally, unless they are disturbed or unless there is a warming trend, as after a rainstorm. They will also bed down near the spring holes if there is an adequate cover of marine vegetation, so there are only so many eels in a given place, and as they are speared out, there are fewer and fewer.

Eel spearing—or "eeling," as it is more commonly called—done either from a small, light skiff, or through a hole chopped in the ice, is not only a way of making a dollar; it is an art. For some of us, it is also a favorite form of winter recreation. It is done, of course, with a spear made especially for the purpose, and a good eel spear is cherished and taken care of more carefully than any other piece of saltwater equipment. A Thomas-built fly rod, or a vom Hofe reel, is a choice piece of equipment, but so is a Kent or a Barrus eel spear. Hand made, of spring steel, these eel spears are not only a pleasure to work with, but they vastly improve your fishing.

An eel spear is a fan-shaped tool. It is built on a drop-forged knife that is about nine inches long. It is blunted at the bottom end, perforated at the top to take three or four U-shaped pairs of tines, and ends either in a threaded tail two inches long, or in a sleeve sort of socket to take the pole. The tines, slightly shorter than the knife, are hand made in pairs and set so that each pair is slightly shorter than the next inside pair. They are tipped with an inward-and-upward-turning barb, an inch or more long. (The barbs tend to get shorter as the years go by, because they are kept filed to needle points.) The tines themselves are barely ¼ inch square at their thickest point, where they go through the knife, tapering to a knuckle-shaped end, which turns back to the barb. The tines are also shaped so the tip of the barb is less than one inch from the next

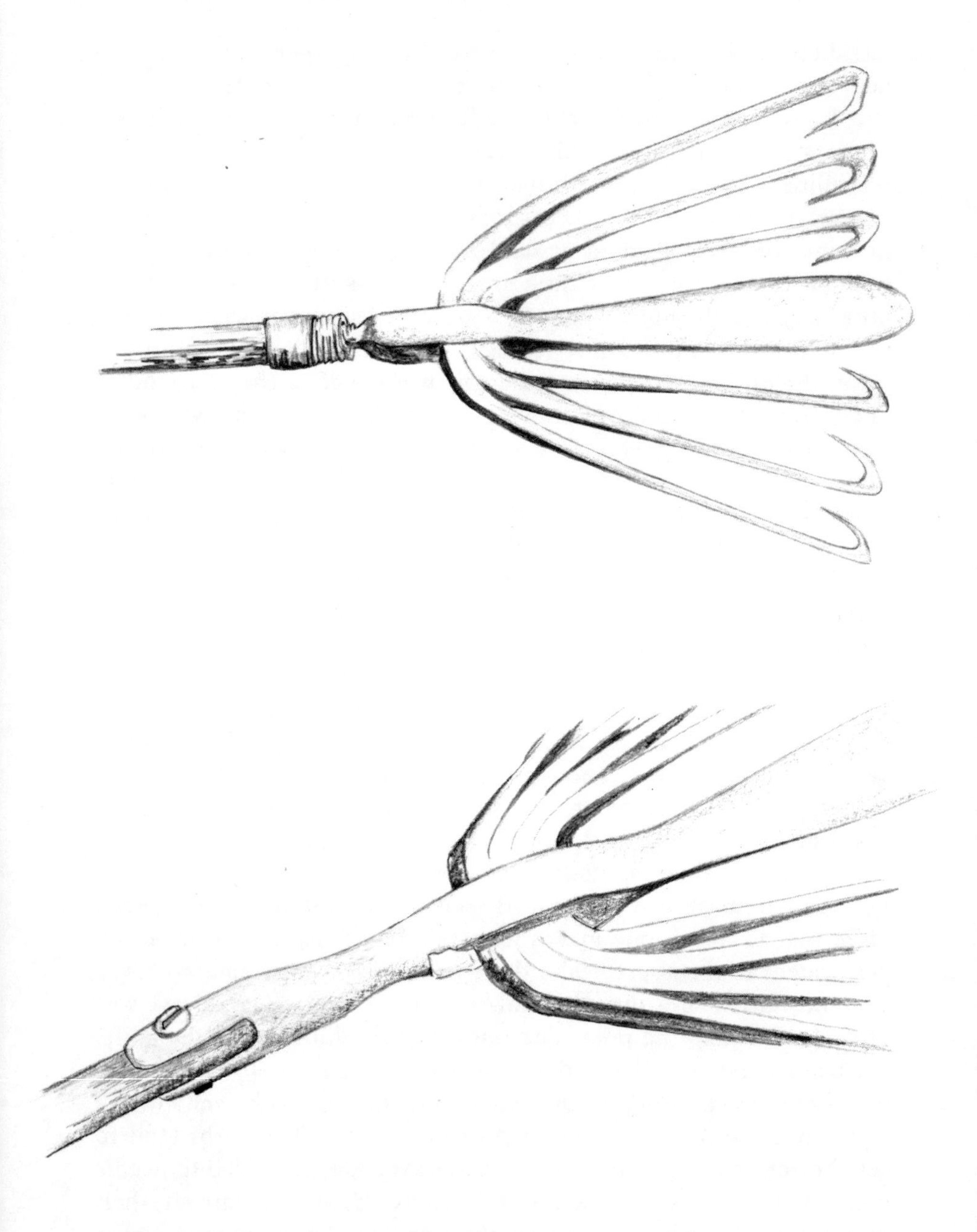

A good eel spear is a cherished tool made of spring steel. Top: a six-tine spear. Bottom: detail of an eight-tine spear with a sleeve attachment.

inside tine. The knife must be stout enough to stand constant jabbing against the bottom, whether mud, sand, or stones. The tines must be limber enough to spring out when jabbed down over an eel as much as two inches in diameter, but with enough spring to come back to position even if the eel turns out to be a two-inch-thick log.

Topping the spear is a pole some 14 to 16 feet long. There are eels in deeper water, but when you spear with a 22-foot pole, which is the longest I ever used, a good many of the speared eels manage to squirm off before they can be yanked to the skiff or onto the ice. Straight-grained long-leaf or hard pine is best for an eel pole. It should be so full of pitch that it will all but sink. In planing it down to size, the fisherman should need a kerosene-soaked rag to clean the pitch from the plane. Substitutes may be fiberglass (and, all too frequently these days, fir, 1 1/16-inch curtain poles). But a good, hard-pine eel pole, not more than 1⅛ inches at the butt and no more than ¾ inch at the top, is what the doctor ordered.

If you're going eeling before the salt ponds freeze over, you will need a good, light, 12- or 14-foot skiff, an anchor rode 16 or more fathoms long, and two light but reliable anchors. And, for goodness sakes, unless you're traveling a long way to the grounds, leave your outboard home and take a pair of oars. Have the anchors ready with the rode coiled down, because you may want to heave over an anchor in a hurry. Push off from shore with the spear, giving it a straight push without any bending of the pole, because it is slender and too much sideways bend can break it. The water may be two feet deep, the bottom hard. Drift and push again, jabbing straight down with the spear. And again, and again, until suddenly you come to softer bottom. You'll be surprised how delicate and revealing the right gear can be. If you don't know your location, you'd best chuck the anchor back toward the way you came, and let run a little scope. Try the bottom with a straight-down jab. Not hard, just enough to penetrate a light layer of mud.

If the mud is not soft enough, let out a few more feet of scope and try again. You won't want soft, sticky mud that holds a foot or more of your spear. There may be eels in it, particularly if it's covered with a thick layer of sea lettuce or spongy wool, but it's hard fishing, and it's better to try that later, when the good eeling is finished.

The eeling jab is done with a fairly slow downward push, and a lightning-quick jerk upward a foot or two. It isn't the downstroke that hooks the eel, it's the upstroke. Push the spear straight down, slowly, alongside the skiff, then pull it up quickly. Wait a couple of seconds so the boat swings a little. Down and up again. You'll know when you've hooked your first eel.

After one of those downstrokes, the spear won't come back easily. Now pull it up quickly, hand over hand, letting the tip of the pole fall where it will. The spear end comes into your lap (did I say you were standing up?), and there's your first eel. If it doesn't squirm off immediately, you may have to roll it farther up onto the spear to get the tine out of it; then it will squirm into the half barrel or keg. Where you found that eel, there may be another, so plunge your spear down and draw it up again, and again, and again.

If you're lucky enough to have found a good little spot (you'll be amazed how much the boat will swing on one anchor), lay down your spear and throw the other anchor as far as you can. (You can't do it if you're standing on the rope.) Throw it slightly across the wind. My former partner taught me the trick of lying in the bight between two anchors. The wind will hold you against them, and you can let yourself downwind by slacking either rope, or pull yourself forward or back as you find eels astern, forward, or off the beam. If you run out of eels to capture, try to follow the same kind of bottom in which you found the first one. If it's hard on one side and soft on the other, you're on an edge, and there are likely to be eels somewhere along it.

Much the same procedure is followed when you spear through the ice. With an axe or ice chisel, cut a hole a foot or so across. Remember, saltwater ice is not the same as freshwater ice: while salt ice must be a little thicker to hold you, it will be softer. Clear your hole of ice chunks and most of the chips. Spear straight down at first. If you find eels, work your way around the hole, working out farther and farther as you clean the eels from the bottom. I find it easier to back up around the hole. In fact, I know only one old-timer who works his way around the hole walking forward, but the eels won't know the difference, whichever way you go.

Follow the bottom edges, hard to soft, much as you would from a skiff, except that you can be much more thorough on ice, cutting your second hole close enough so that you can hit at least the fringe of where you worked the first hole.

When spearing through the ice, you ought to remember two things: (1) Salt ice is apt to be tricky, and when you are working on it, it's good to have a partner along. He may not be able to help you if you fall through and drown, but at least he can tell your widow what happened to you. (2) Keep track of which way the tide is going: more than one eeler has found that the flood tide has come up between him and the shore, and he has been marooned longer than he wanted to stay afloat.

A tip or two about shipping live eels: If you plan to ship live eels in the summer or while the weather is still warm, buy a new barrel and make sure the bottom is nailed tight. Place upright in the barrel a chunk of ice, preferably the width of a 300-pound block, and a foot square. Pour the live eels around this, but prepare yourself for an explosion of eels over the top. The ice seems to "burn" them, and it makes the eels unhappy. When the barrel is as full as possible, cover it with a double thickness of burlap bag, and secure the bag by driving down the top hoop and nailing it tight. If you ship eels during cold weather, the cake of ice is not altogether necessary. If the fish house from which you ship has no barrels, winter eels can be shipped in a "black back," or flounder box. Be sure the bottom is nailed in three or four places on each side, then put in a skim of cracked ice, then your eels, then more ice, of course. And nail the top much more tightly than you would for cod or haddock. Skun (or dressed) eels may be shipped in the same kind of container, but don't just dump them into the box in a snarl. Straighten them out as best you can, and they'll bring a higher price.

EEL SKINNING

Very few eelers bother to save eelskins, but there are times when salted winter eelskins are worth more than the eels themselves, especially when the stripers and bluefish run as heavily as they have for the last few years. I never enjoyed skinning only one or two eels: you get just as slimed up as though you had skun a kegful.

You need a very, *very* sharp knife, and a stone to keep it that way. No fish-cleaning operation takes the edge off a knife more than eelskins. At shoulder height, drive an eightpenny nail, slanting up, into a wall. Let it protrude about an inch. Then cut off the nail head.

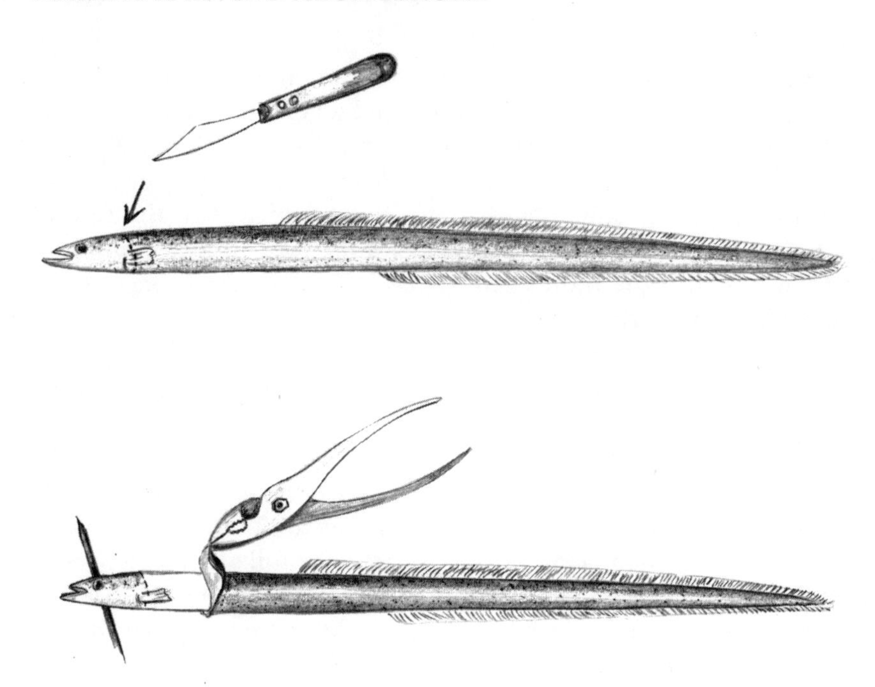

Eel-skinning is a two-step operation.

Hold your first eel just back of the ears (the pectoral fins). Circle it with the very sharp knife, cutting just deep enough to cut through the skin, but not deep enough to cut into the meat. Impale the eel on the nail at this slit. (If you're cleaning eels simply for the meat, slit it from vent to neck; otherwise leave the belly uncut.) Take a good pair of pliers and, nipping just the edge of the skin, try to peel off the skin, much as you would turn a stocking inside out. You may have to start peeling in two or three places, because you don't want to tear the skin, nor do you want any of the meat to adhere to the skin (the meat will turn brown when salted). Once you're started, you'll have no trouble until you come to the vent. If you pull the skin carefully by this point, and the skin begins to tear, cut lightly crosswise just astern the vent, and then continue turning the skin inside out. The more the skin is split, the less value it has to the troller. To finish cleaning the carcass, slit the belly, strip out the very small gut with your thumb, and cut off the head.

The skins should be dry-salted, not kept in brine (although

their own fat makes a brine with the salt). A five-gallon crock is ideal. Cover the bottom of the crock with coarse-fine fish salt. (If it's not available, kosher salt is the next best. You can use ice-cream salt or ordinary table salt, but the former is too coarse and the latter is too fine.) Put a layer of eelskins on top of the layer of salt, then put in another layer of salt, and so on, until you've filled the crock or run out of eels. Cover well with salt, and keep the crock covered, out of the sun, and in a cool place.

The best market for dressed eels probably will be the local fish market, but I have iced them, shipped them in a "flounder box" to New York, and done very well.

Unless you have kept 12- to 14-inch eels for live bait (and turned them loose—the unsold ones, that is—at the end of the season), don't keep anything under ¾ pound. You'll spoil your own price if you sell them alive, and if you're skinning them, it's hell to wind up with 30 or 40 little eels in the bottom of the keg. They're as much trouble to skin as the big ones, and they don't add a thing to your weight.

4 CLAM DIGGING

Digging for soft-shell clams (steamers, long-neck clams, piss clams, *Mya arenaria*) by hand may not be the easiest way to make a living along the shore, but it is one of the easiest skills to learn, provided your back is strong enough, and one of the least expensive projects to undertake. After you have acquired your shellfish license, the only equipment you will need is a couple of clam hoes (clam diggers, clam forks, or clam hacks, depending on where you live) and a means of transporting your clams, whether in baskets or hods (or "drainers," to use the old phrase. "Dreeners" is the way the local ancients said it.). Of course, you will need to work in an area affected neither by pollution nor by *Gonyaulax tamarensis* (the "red tide").

You will probably need a couple of clam hoes (or more), because not all the bottom in which you'll find clams is the same. One type of hoe works in firm sand, another in sticky mud, a third in wet, porous sand. For some reason, not excepting the so-called Ipswich digger, there is no good clam hoe available in any local hardware store. The teeth on "boughten" hoes are too short or too blunt, the angle where the tang fits the handle is either too flat or (as with the above-mentioned Ipswich hoe) too steep, and the handle is far too long. Except for the fact that they cost too much, there isn't much else wrong with them.

If you are planning to dig in coarse, watery sand where the clams are relatively deep (and while the biologists may say *Mya* is

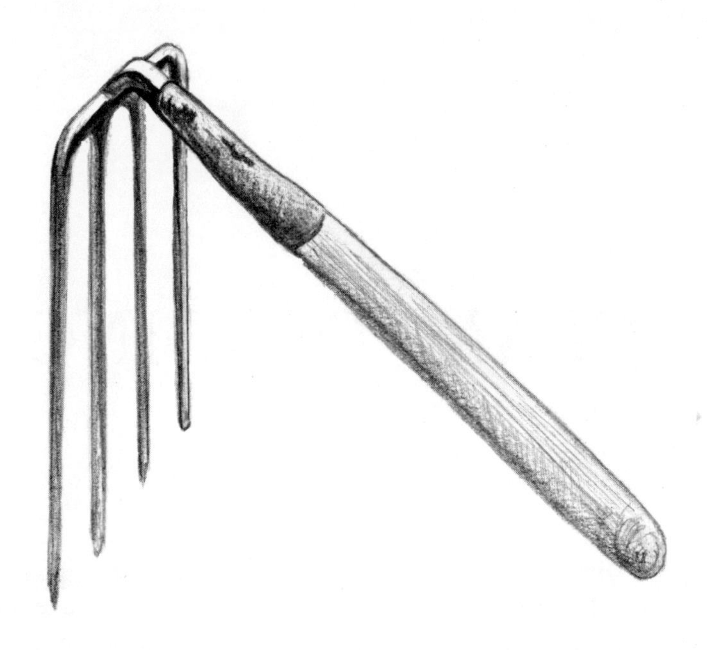

The so-called Ipswich digger. Theoretically, the excessively sharp angle between tines and handle helps the tool slide off coarse gravel and stones.

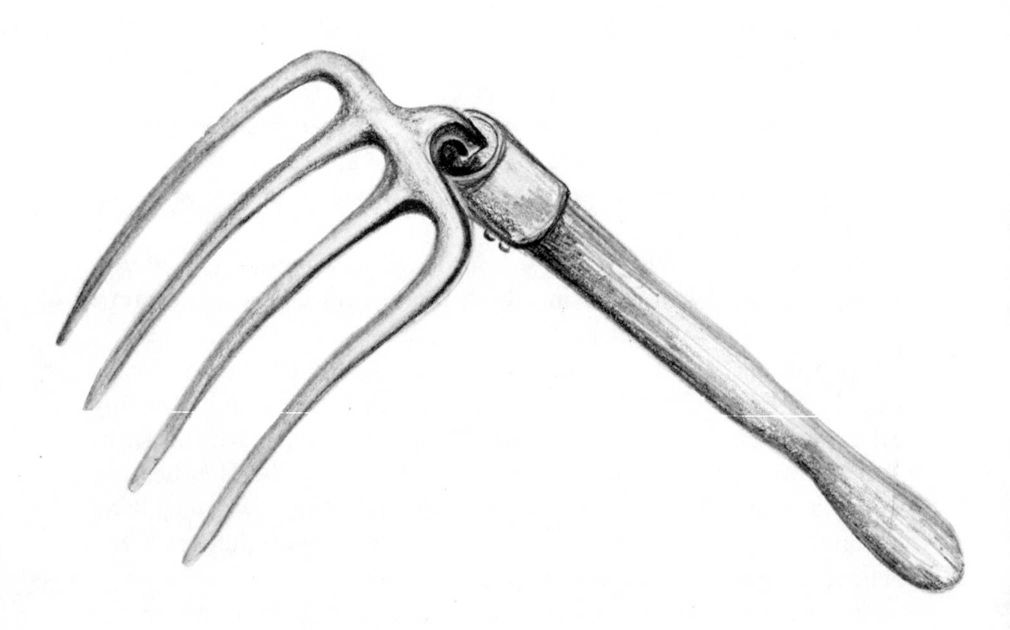

The author's favorite clam hoe.

found two and a half times as deep as its own length, this is variable), you will need a four-tined hoe, with a good "hook" in the teeth. The teeth should be wide (perhaps as much as ¾ inch) and long (at least 10 inches long and perhaps as much as 12 inches in extreme cases). The handle should certainly be no longer than 18 inches. I prefer 16 inches.

While this hoe can be used in wet, soupy sand or mud, it would kill you in tightly packed sand or in mud with a large clay content. In damp but not watery sand, the clams probably won't be as deep. My favorite hoe for this kind of digging was made of an old, "ladies" spading fork, with teeth ⅜ inch wide and 10 inches long. If you are digging in clayey mud, the clams probably will be even shallower, so a six-tined hoe with rounded tines eight inches long is a better tool. Since none of these tools is available locally, the local men resort first to the hardware store, then to the blacksmith. If there is no local blacksmith, most garages have a mechanic pretty well skilled with a torch or electric welding outfit.

Buy a spading fork (pitchfork) with tines that most nearly fit your needs. Have the blacksmith heat the tang of the fork and bend it so it is at slightly more than a right angle to the tines. Saw off the handle to the length desired. Grind down the tines at the points, making them not narrow but flat, being careful not to burn

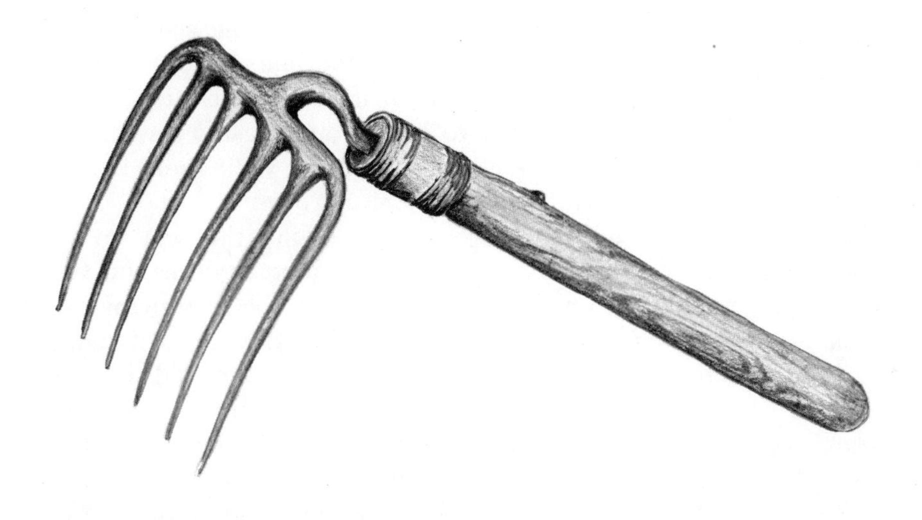

A hard mud clam hoe.

them and draw the temper. They don't have to be ground down to a knife edge, but remember that the thinner they are—up to an inch or so above the points—the more they will cut into the mud or sand. A good clam hoe, shaped to fit, is a tool no regular clammer allows anyone else to put his hand to carelessly. You don't have to buy a whole set of clam hoes, so to speak, at once. Buy the one most suited to your present needs, and then keep your eyes open for secondhand spading forks. The old ones seem to have better steel and better shapes.

Now about the actual digging of clams. I expect that half the commercial diggers on the coast will laugh at me, but I have dug my share of three-barrel days, and I'm presuming you are new at the business and want to know what to look for.

Let's take optimum conditions—a bar of firm, not-too-wet sand—and go from there. The tide is dropping and the clam holes are showing, indiscriminate holes, round. But you will notice that the holes are in clusters—a bunch here, a gap there, then a long stretch where the holes are reasonably close together. Figure out in your mind's eye where you are going to dig. Don't start with the first thick clump of holes; start back a foot or two in clear bottom. Make your hole three or four forkfuls wide at first. Take the top off the sand the first time across, then go deeper the next time, and just before you get to that thick fringe of clam holes, dig down until you are sure you are deeper than the clams. Take out a couple of forkfuls of sand so that the hole is deep enough and wide enough for you to put a boot inside on each side of the hole. That way, you won't have to bend your back so far, and if you go down on your knees—as you'll probably want to do before the tide is over—you're going to slow yourself down.

Now, push down your digger teeth beyond the first lot of holes. Hook the forkful back, using slightly more pressure on the top of the teeth than on the bottom. (This is the reason for the hook in the hoe's teeth: to make it hang in.) If you do it right, the sand will topple over into the hole and the clams will be left bottoms up. However, if you don't go deep enough the first time, scratch out any clams you can. Clean out the area where you have dug, take off a half-forkful of sand and toss it between your legs, and then go down a full forkful. You'll find that most of the clams you want, the ones that are two inches long or more, are nearly all at the same level. Once you establish where that level is, you'll know just how deep to dig. The small ones you don't want will go out between your legs, and the big ones will be still deeper. Unless the digging is sparse or the market wants them, don't bother with the big ones. Each time you go ahead, clean your hole out slightly deeper than the bulk of the clams. Follow the inshore edge of the thickest clumps of clams, carrying your ditch, so to speak, with you. If the tide is out, you may want to follow the edge of the bed all around (the clams will be bigger where there are fewer of them), or you may want to double back alongside your first strip. All this time, your drainer has been kept near your left hand, hitched slightly forward each time you moved ahead.

Of course, there are infinite variations on this procedure. If the sand is so porous that the water runs in where you are digging, you'll want a hoe with wider teeth and you will have to move faster. You can't wait too long, or the water will get ahead of you, and you'll be trying to wash out your clams. If you are in hard clay-mud, water may not be a problem, but it may take a lot more backbone to turn the clams you want upside-down. In compact bottom, two-inch clams usually won't be deeper than the length of your clam hoe's teeth. If you get in sloppy mud, particularly if you're digging big clams, you may have to carry an extra bucket to use in bailing out the excess water.

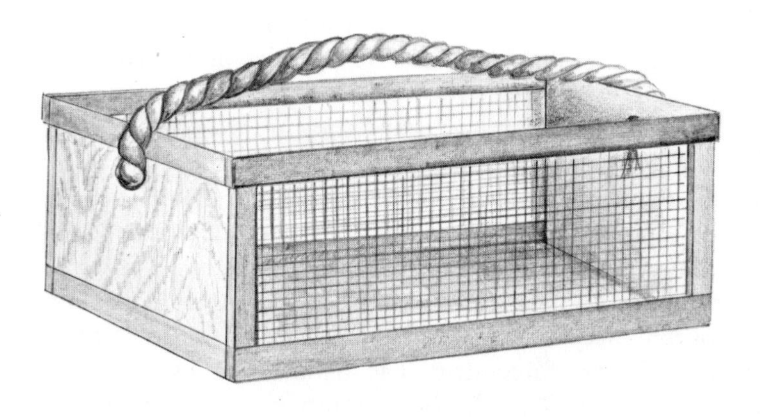

A standard clam hod.

Dyking clams is a technique for the experts. It is usually done in soft mud bottom where the clams are never uncovered at low tide. It's not the easiest digging in the world, but it can pay off when you learn it. Start inshore, but instead of throwing the mud out between your legs, throw it ahead of yourself, to build a dam or dyke around the clams you want to dig. Each time, throw the mud ahead until the dam is higher than the water surrounding it. As you dig to your dyke, push the mud ahead, building the wall farther and farther out into the tide. Once in a while, if the water gets ahead of you, or if your dyke leaks, you may have to bail yourself out with that extra bucket.

When you are finished, don't leave a lot of juvenile clams exposed. Take the time to fill in the last of your digging, covering whatever short clams would be exposed to too much sun or to the hungry mouths of minnow fry when they come in on the flood tide.

The modern trend seems to be to put clams in baskets as they are dug, but I still think the old way was better. When the weather was bad and we had nothing else to do, we used to build clam hods, a few more than we actually needed. In the days of three-barrel digging, when I started, that meant a lot of hods. The inside dimensions of each hod were nine by nine by 19 inches, and each held two 12-quart buckets of clams, or two-thirds of a bushel. We filled five hods for each three-bushel barrel, with the extra bucket for shrinkage. Today, even though so many clams are being sold by the pound, the hod is still a good container, because it can be set on the edge of the tide line, just underwater. The clams are washed, kept cool, and to a certain extent left to spew out the sand or mud that they inhale while they are being dug.

Standard hods (and there were a great many that did not conform to the standard, of course) had two square ends (usually made of scrap lumber). The hods themselves were made either of second hand laths set a lath's thickness apart and cut 20½ inches long (to allow for the ¾-inch-thick ends), or, in later years, they were covered with cellar window wire and battened at the corners with laths for reinforcement. The handle was usually a piece of ¾-inch rope secured on the outside just above the middle of each hod end. Or it might have been a thinner piece of rope threaded through a short length of old garden hose so the rope wouldn't cut through your hands if you had to struggle across a hundred yards of sticky mud with a hod full of clams in each hand. The clams can be rocked gently underwater in these hods, until they settle, and then they can be rocked with a slight up-tossing motion from end to end until all the mud or sand has been washed out.

Half-bushel wire baskets are good if you can afford them. However, not only do they cost more (and I don't wish to impugn anyone's honesty), but they also have a way of disappearing if they are left around the shore.

You have your gear and your clams; now you need a market. Fishermen generally forget that the better condition their catch is in when it reaches the market, the surer they are to receive top price. If a fisherman consistently brings in clams that are clean of mud and sand, if there is minimum breakage, if "short" clams are never brought in, then on the days when the market is slack, his clams will be the ones the fish buyer wants.

I hold no love for fish buyers, having fought with too many in

The author dyking for clams.

my day, but you have to remember that they are in the business to make money, too. There is another angle from the fish buyer's point of view: he has to be able to depend on you. He has had to fight the fish market for a place to sell your clams, and, in all probability, he had to promise to deliver a specified amount of clams at specified times. If you have a hangover, or if the tide or weather is phew, or if you feel like a holiday, the fish market couldn't care less, and the fish buyer is caught in a bind. Guess whose clams he is going to buy the next time there are more clams available than there is a market for? It seems fundamental, but think about it.

Of course, if you're lucky enough to con the local fish market into buying your clams, thereby cutting out the fish buyer, the same rules pertain: clean clams; no broken or short clams; and the order filled, fair weather or foul. Keep your clams cool until they get to market. Don't carry them in buckets of water: the water will heat up before the clams do, and the clams will use up all the oxygen in the water and drown. Keep them out of the hot sun, covering them with a wet burlap bag if necessary.

5 QUAHOGING

The shellfish called a quahog on Cape Cod is called elsewhere a round clam, hard clam, chowder clam, cherrystone clam, littleneck clam, *Mercenaria mercenaria,* or, to use the older term, *Venus mercenaria.* Littlenecks and cherrystones, in our area, are simply juvenile quahogs, two or three years old, which have not yet grown to the size of "sharps" or "chowders." In Massachusetts, the law allows no quahogs to be taken that are less than 2 inches across at their widest diameter, whereas in Rhode Island and Connecticut the minimum size is 1½ inches. The various state laws, then, govern the minimum size of littlenecks. The commercial market, however, controls arbitrarily the maximum size. With a buyers' market, the Massachusetts commercial fisherman may be limited to littlenecks that are no more than 2¼ inches long. On the other hand, if there are few littlenecks for sale and the demand is strong, then the buyers may say, "We'll take anything up to 2¾ inches as littlenecks." Cherrystones run seasonally from the upper size limit of littlenecks up to three or even 3¼ inches long, again depending on supply and demand. Usually, anything over three inches long is considered a sharp or chowder quahog.

In a good many coastal areas, quahogs are the bread and butter of the shore fisherman, the "you-can-always-make-a-buck" shellfish. This is partly because the market is relatively stable, but more importantly, it is because quahogs grow in such a wide variety of conditions, anywhere from intertidal zones out into deep water, 30 feet or more, deep enough to require power machinery. They grow in bottom ranging from coarse gravel and rocks to the silty mud of slough holes in a marsh. Their shells are hard, and the shellfish will stand a lot of abuse in handling. They grow in so many different places, and under such a variety of conditions, that the methods and machinery of harvesting vary widely. Since this book takes in only the hand work that is possible with a small skiff, we'll neglect the power dredges, which vary from the hydraulic dredge to the New Bedford "rocking chair" to the sand "knife" dredge. If you feel you have to get into that business yourself, get yourself a job culling on a good working dragger for a couple of

A chowder clam, with relative dimensions for other sizes of quahogs.

years, and even then you'll find you haven't learned all you need to know.

We shall consider, then, only three ways of quahoging: hand scratching, tonging, and bullraking.

SCRATCHING QUAHOGS

Scratching quahogs—backed up by holing, treading, and turfing—requires the least expenditure of capital of any method of earning a living alongshore, except possibly clam digging. All that is needed is a scratcher, a six-tined piece of equipment, with a five-foot handle, that closely resembles a hand potato digger. You will also need a bucket or basket, a pair of sneakers in the summer, and a pair of hip boots or chest waders in the winter. (Also a local shellfish license, of course, plus a tide calendar, lots of time, and a number of burlap, net, or plastic bags—usually furnished by the fish buyer once you've become established.) In some areas, you don't even need a skiff for quahog scratching. This is a low-tide proposition, and the lower the tide the better, because it lets you get to areas farther offshore than usual and gives you more time to work.

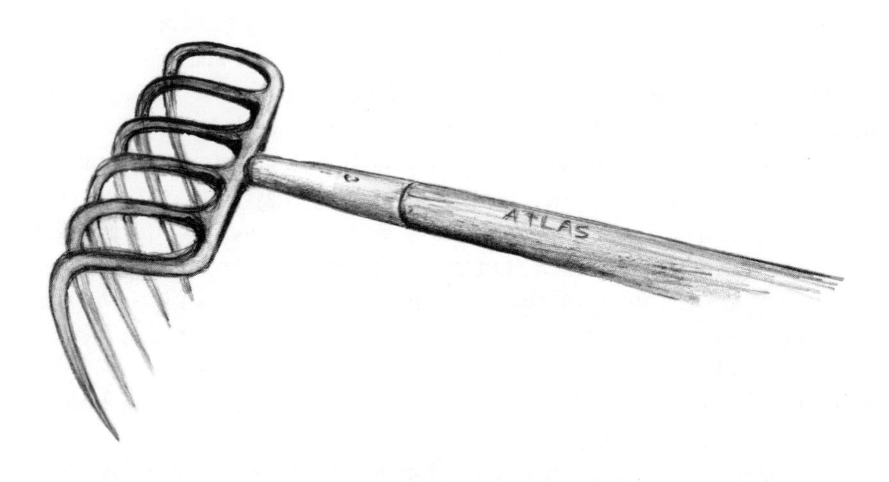

A hand scratcher for quahogs.

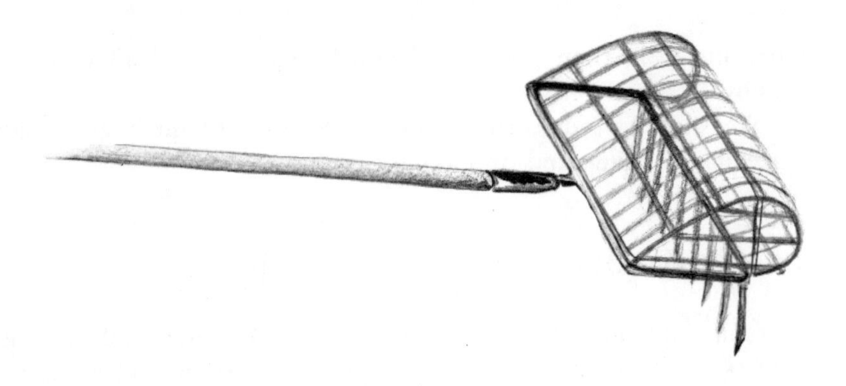

The basket scratcher for quahogs. It is sold only to tourists and is good only in clear mud or soft mud bottom. It is useless for scratching quahogs in stony or shelly bottom, particularly in the winter, when quahogs are deep.

Holing quahogs is perhaps the simplest (after you have learned it) and yet the most sophisticated way of taking quahogs. Holing is a trick you have to learn. Quahog holes show best in the spring and early summer, because quahogs are more active then. In cold weather, quahog holes seldom show at all. In some areas, they show best on a light east wind, during a warming trend in the weather. The holes are made when the quahogs stick out their siphons to feed. In dry sand, watch for a slight splatter that looks as though someone had dropped a teaspoonful of water from a couple of feet up. If the sand is firm, you may see a keyhole-shaped hole in the sand, up-wind of the splatter. In black, hard mud, there may be the same keyhole surrounded by a circle of lighter-colored mud. At low tide, in slough holes in a marsh, where the water has not completely run off, you may actually see the siphon at work. It may appear as a miniature yellow dumbbell, perhaps rimmed with black. On no two days and in no two places do quahog holes look exactly the same. An amateur can go over an area, pecking at every hole in sight, and an old-timer can come along behind him and find more quahogs than the beginner can. You get to sense which are quahog holes and which are clam holes, worm holes, or whatever.

Turfing for quahogs is a neglected method used chiefly by old-timers to search for littlenecks and cherrystones. It is done most effectively on a hard mud bar, and sometimes even where clusters of blue mussels make scratching or holing impossible.

Use an eight-tined clam hoe. Once you've started a hole three or four hoes wide and two or three inches deep, move ahead about six inches; bang the hoe teeth into the mud and hook back with a sort of tip-over flip. You should have dug deep enough to half-uncover the young quahogs. You can use one hand with the hoe and the other hand to pick up the quahogs. Occasionally, you might accidentally stick the tines through the back of a young quahog, but if the quahog holes aren't showing, that can't be helped.

When you get a little farther offshore, where the tide has still not left the flats, treading begins. (You may want to go barefoot when treading, but doing so usually results in cut feet, and you'll be amazed at how sensitive your sneaker soles can become.) Quahogs are found only slightly below the surface, so if you are walking on them in bottom that is anywhere near soft, you'll feel them as hard lumps, which can be dug up either with your fingers or with a scratcher.

Then you can begin to scratch or "rake." The scratcher tines dragged through the sand or mud will "bounce" when they hit a quahog. A second, deeper scratch will bring your prey to the surface. Most of the year you don't need to go deep; an inch or so will hit all the quahogs. But in the winter you may find quahogs much deeper, so much so that sometimes it pays to make a second scratch where you made the first one. Other than picking up your catch and lugging it ashore, that's all there is to it, except that, to be commercially profitable, quahogs have to be fairly plentiful. If you don't find a quahog every other scratch or so, keep moving. The one sure difference between a commercial quahoger and an amateur is that, after finding a single quahog, an amateur will stay in the same place, patiently working over every inch of the bottom until the sun goes down or the tide comes in; while the pro, knowing his time is limited, will scratch and move, scratch and move, roughly covering a great deal of ground until he hits pay dirt.

The one other single piece of equipment that will increase your speed is a scratch bag or drag net. This can be made of netting fine enough so no littleneck will go through, or it can be simply a burlap sack tied on opposite sides to a loop of rope long enough to go over your shoulder. A simple refinement can be a hoop, 10 or 12 inches in diameter, whipped to the mouth of the sack to keep it open. This rig will increase your productivity immeasurably if you are scratching in water deeper than your pail or basket, and it has the added advantage of leaving both your hands free.

TONGING

The use of tongs for quahoging is strictly limited in the Cape Cod area. To be practical, they can be used only in relatively soft

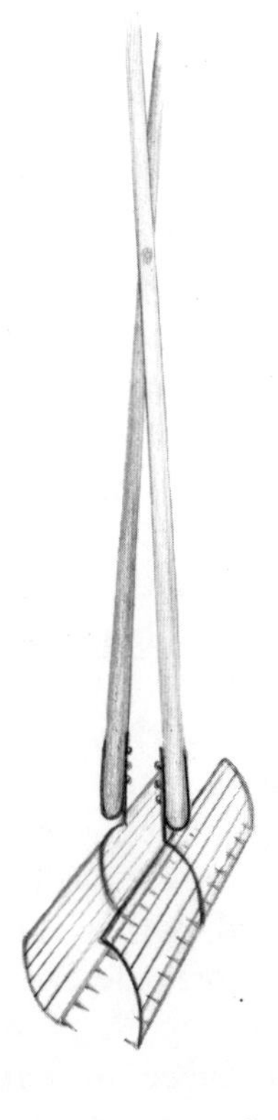

Tongs are used occasionally for quahogs, but they are best left to oyster fishing, for which they were invented.

(mud) bottom, in water from wading depth (three feet) to no more than 12 feet deep.

Imagine a gigantic pair of shears used by the wrong end. They are operated from what would be the cutting end. Instead of the circular guards for thumb and forefinger, there are opposed and matching semiflat baskets, which are toothed on the lower edges. The long parts of the shears are matching wooden "stiles" or handles. From talking with the men who use them (to be quite honest, I never used them successfully, partly because I tried them in "iron" bottom—among rocks—and partly because the quahogs I was after ran from 90 to 140 to the 90-pound bushel), I gather that 16-foot stiles are the maximum.

You must anchor your skiff tightly, fore and aft. Put down the toothed baskets, the working end of the tongs, and spread the top ends of the stiles as wide as your arms can reach. With the teeth digging into the sand, work the tops of the stiles with a down-jabbing, squeezing, lifting cycle. Spread the top ends of the stiles and repeat. Bring the stiles up, squeezed shut and with the baskets horizontal, and pick out your catch. This gadget was designed for catching oysters, which lie on the surface of the bottom, so why not give it back to the oysterman and go to bullraking?

BULLRAKING

I have no idea who first invented the bullrake. It probably wasn't invented but it evolved, developed by hungry fishermen who wanted better ways of going after quahogs in deep water. *A Report upon the Mollusk Fisheries of Massachusetts,* published by Wright and Potter Printing Co., State Printers, in 1909, is a fascinating and amazingly modern work that includes pictures of bullrakes, both "basket" and "claw" types (the latter is also called a "Nantucket" rake). Bullrakes have changed little in the nearly 70 years since then. Whatever the type, whatever the local changes in tooth position, shape, and size, the bullrake is fundamentally a toothed tool with a basket, which is either attached or an integral part thereof and fastened to the bottom end of a long pole. The whole rig is thrown overboard, teeth down, to the limit of the pole.

It is jigged or worked along the bottom, toward the boat, and twitched out and brought up to the surface, teeth up.

Let's take the rakes one at a time. The "hard-bottom" rake may have anywhere from 12 to 16 teeth. The teeth may be from one to three inches long, set from an inch to 1⅝ inches apart. Currently, locally, this rake costs about $2.50 a tooth. The frame—a rectangle as wide as necessary for the number and spacing of the teeth, and perhaps a foot deep—may be made of ½-inch- to ¾-inch-square iron. It is finished off at the exact center of the back of the rectangle with a tang, an eight- to 10-inch extension at right angles to the frame, ending in an upturned tooth that bites into or is clamped to the pole. Three or four rounded bows are welded on the back or upper side of the frame. Heavy twine, chicken wire, or welded steel straps are secured to the bows to form a basket when the rake is turned teeth up.

This contrivance is secured to the end of a pole, which can vary in length from 12 feet to as much as 30-odd feet. (I have used a 55-foot pole, but there are stories of old-timers using 65- and even

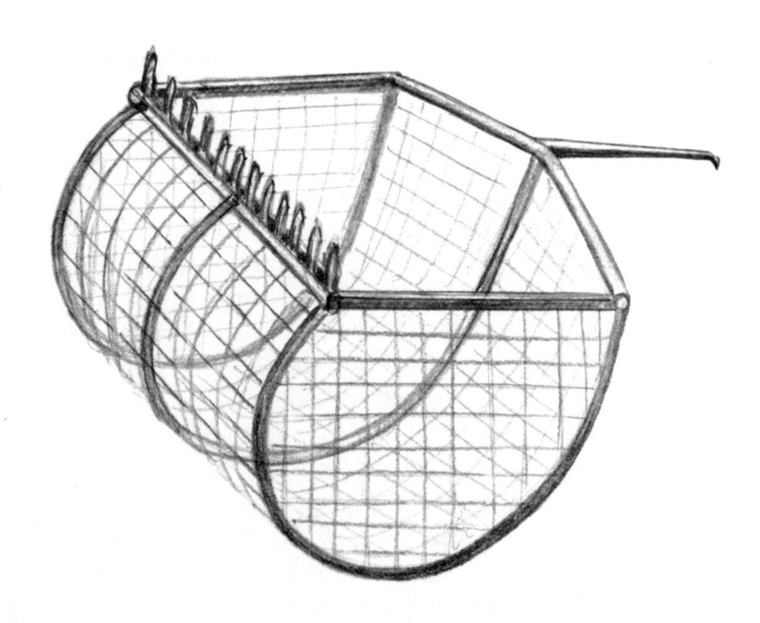

Hard-bottom quahog bullrake. Used in water eight to 20 feet deep.

90-foot poles.) Today, poles seldom reach more than 30 feet. The poles traditionally were made of tapered hard pine, perhaps 1⅜ inches thick at the bottom and ¾ inch thick at the top. Today, telescopic poles of aluminum tubing have appeared, with two sections making perhaps a 20-foot pole, or three sections stretching almost to 30 feet. Locally, they cost about $1.00 per foot. Having learned the trade with hard pine, I feel the old-timers had the better system. They had a rhythm that let the pole do most of the work, which I haven't noticed in the rakers working with aluminum poles (with a "tee" crossbar at the top end). But long-leaf hard-pine poles in 30-foot lengths are, I'll grant, difficult to come by in these parts, so perhaps the bullrakers are making the best of a bad time.

There is a variation in the hard-bottom rake that is now quite popular locally. It has slightly hooked teeth that curve from the outer crossbar toward the handle end. The teeth, made of flat steel, vary in length and curvature, but they are approximately two inches long. This type was known years ago, logically, as the "eagle claw."

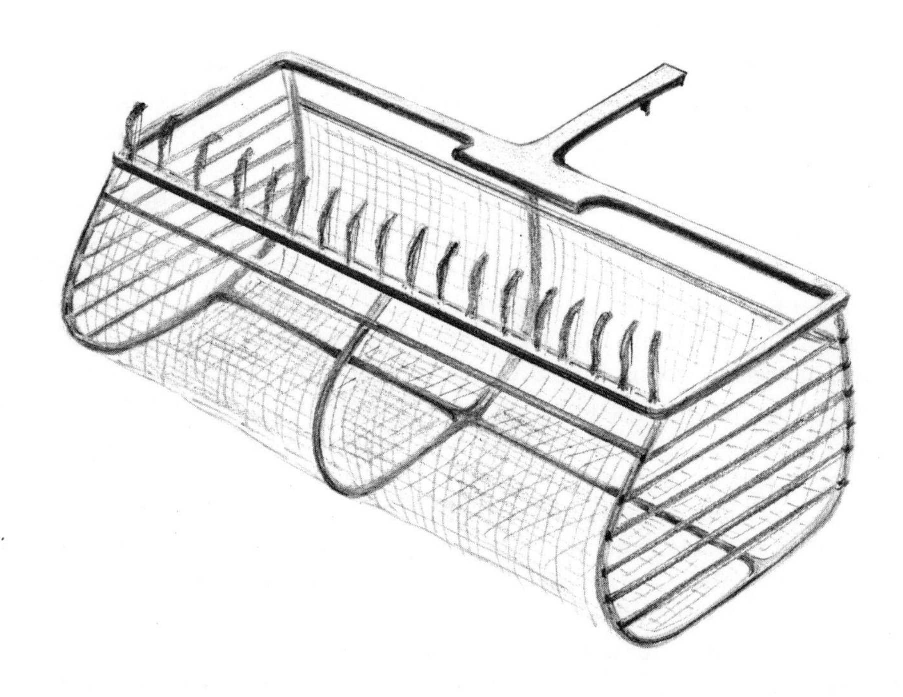

A modified eagle-claw rake. Used in medium hard bottom.

The mud rake has been developed more, perhaps, in the last 10 years than any other type of quahoging tool (if you can exclude the switch from hard-pine to aluminum-tubing poles). The original mud rake was very much like a miniature, horse-drawn hay rake. It was perhaps 36 inches wide, with curved teeth set no more than an inch apart. The teeth were welded to an iron bar across the back (to which was welded, at a slightly different angle, a tang very much like the tang on a hard-bottom rake). Most of the original mud rakes had round, ⅜-inch steel teeth, flattened and pointed at the working end and curved slightly more than halfway around a circle a foot or more in diameter.

The modern version of the mud rake has extended the flattened teeth as much as 10 inches in some cases. The teeth are stiffened by a bar welded on the outside or back side of the teeth where the flattening stops and the round steel begins. The circle has been reduced to as little as six inches, making a sort of oval basket. To compensate for the longer teeth, the tang also has been cocked slightly upward.

Whatever type of rake you use (and it is best to consult the local fishermen regarding any minute variations, since bullrakes are developed for peculiar local conditions), and whatever length and kind of pole you use (again, consult the local fishermen for specifics), the raking sequence is nearly standard.

If you are new at the game, anchor your flat-bottomed skiff by the stern. Run out the anchor rode up into the wind, or nearly so, the full length. You'll find yourself happier with an extra-long rode than with a short one. I'd settle for 30 fathoms if I were raking in two fathoms of water. Three-eighths-inch "poly" or nylon rope is good. Flip over the bow anchor and drop back eight or 10 fathoms, depending on the kind of bottom you are fishing and the type of anchor you have. It is better to allow a little extra rode forward than too little and then have the bow anchor pull out just when you get into good raking. With blacksmithing being what it is today, and the cost of anchors exorbitant, the sizes and types of anchors may vary. Had I my "druthers," I'd rather have at least a 10-pound sand anchor aft, and anywhere from a 12- to an 18-pound sand anchor forward. Better an extra-heavy anchor than one that pulls out on you just when you begin to get quahogs in paying loads. If you are going to rake in a muddy area, a wide-

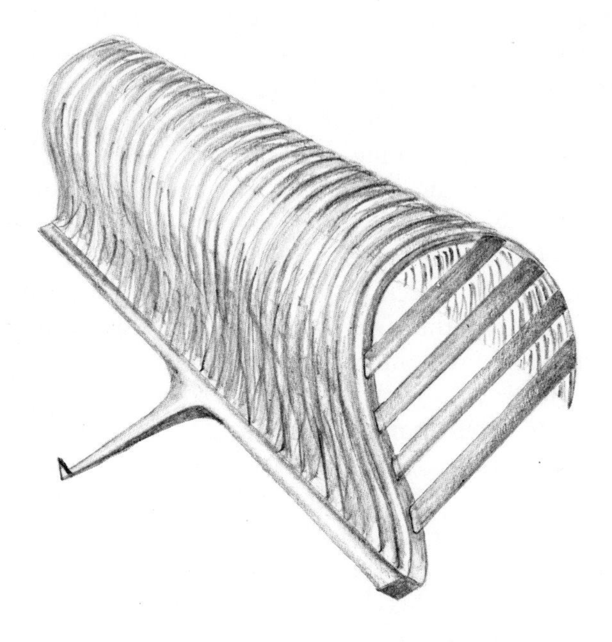

The mud or Nantucket rake. Used only for raking quahogs in soft mud.

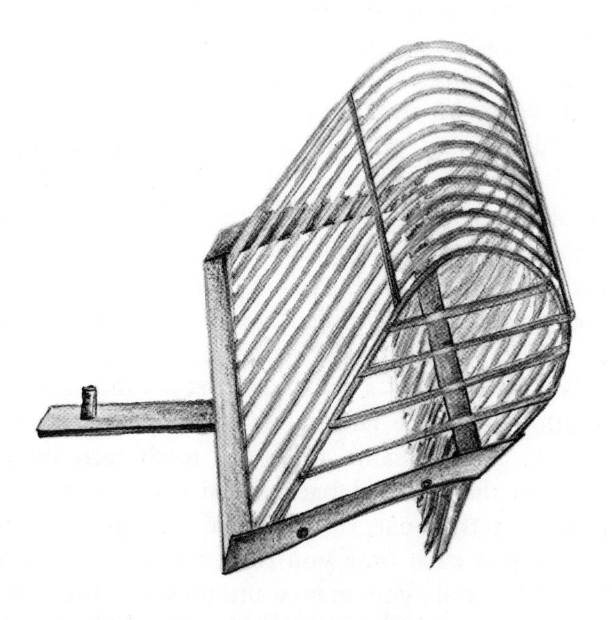

Modern version of the mud rake.

fluked Cape Ann anchor might be better for both bow and stern.

Once the anchors are set and you've pulled down tight on them, both fore and aft (you should have an adequate cleat near the stern on each rail), you are ready to rake. Pick up the rake, with the teeth down, making sure the pole will run clear (with or without a "tee" at the top end). The rake and pole should be balanced before you throw the rake out. The longer and heavier the pole, the farther back the balance point is from the rake end. With an underhand "chuck," throw the rake straight off the stern, and let the pole run through your fingers. If this technique is used properly, the rake will hit the water teeth down, squarely, and your initial chuck will be strong enough so that the top end of the pole will be in your hand before or by the time the rake has settled to the bottom. Now, the trick is to get the teeth of the rake deep enough in the sand or mud so they will rootle out the quahogs, but not so deep that the rake will "bury" itself, and all you will get will be a rake full of mud and sand.

Pull and relax. (This is why I like the old hard-pine poles better: there is a rhythm no one can teach you; you have to learn.) Each time you pull, the rake should come ahead a little; each time you relax, it should dig a little. You should learn very shortly just how hard you can pull the rake ahead without pulling the teeth out of ground. If you're using hard pine, as you recover your pole, let it run up over your right shoulder, and press down with your hands as you pull. If you're using an aluminum pole with a "tee," when you've worked in enough pole, loosen the clamp, telescope your pole, and tighten the clamp again. Whichever gear you use, the vertical swing and bounce of the pole will pull the teeth ahead and set them deeper.

Once the pole has been pulled ahead to the point where it is getting short (with the teeth picking up all the quahogs along the way), there is a twitch and a twist that breaks the rake out of the bottom smoothly, without jarring off any quahogs that still may be on the teeth. At the same time, the teeth turn smoothly upward. The pole is then hauled back, hand over hand. If there is a fork in the bow of the boat, run the pole through it so that a wet pole won't soak you each time you reset the rake. If you're raking in mud or muddy sand, you may want to scrub the rake up and down a couple of times before culling. No matter how many times

I say it, remember that the cleaner your stock, the easier it is to cull, and the better price it will bring. Cull out the keepers—littlenecks in one basket, cherrystones in another, and sharps in a third. Throw the rest (the trash) overboard.

I'll explain the reason for starting at the bow anchor and working back as you clean out an area. Too many times I have seen fishermen do what appears to be the easy way—drop over the bow anchor, drift back astern, then rake their way forward to the bow anchor, all the time picking up the trash and shack that they threw overboard from the last haul.

If the quahogs are good, you may want to make four or five rakes to a berth before you shift back on your rode. If they come very well, you may want to make eight or 10 rakes. If they come poorly, make a rake amidships, maybe one to port, and one to starboard, and then drop back. You can gauge best how far you are raking. If you can make short rakes profitably, drop back a very short distance. If you're making long rakes unprofitably, drop back farther each time. This is the sort of judgment that culls the men from the boys. If you find quahogs at the extreme starboard side of any berth, before you shift anchors, slack off slightly on the stern anchor and scull to starboard, making a throw with the rake before you have drifted back. Again, if the rake is good, you may want to pull ahead on the bow anchor and shift slightly to starboard. You are working blind, and what comes up in the rake is the only indication of which way you should shift. As with almost all fish and shellfish, the best fishing is almost always on an edge, a transitional area between hard and soft bottom. One other thing: it's all but impossible to rake downhill. If you find yourself on the edge of a hole and the wind is not blowing too hard, shift your position so that you are raking uphill.

The value of a third anchor off the beam (usually a throwing grapnel) is debatable unless you're doing "very fine" fishing. Under most conditions, I consider that its helpfulness is outweighed by the complications involved in setting it.

Cull the quahogs clean, cull them honestly to size, keep them cool in summer and unfrozen in winter, and you should get the top prices.

6 SCALLOPING

"The scallop fishery has existed for years, but it did not become of commercial importance in Massachusetts until 1872. At that time there was hardly any demand for scallops. Since then, the market has increased until the supply can scarcely meet the demand. It seems incredible that people should have once looked upon this highly colored shellfish with beautiful shell, as poisonous and unfit for the table, in the same manner as our country forefathers considered the 'love apple,' now the tomato, as only an ornament for the garden. . . ." (From *The Scallop Fishery of Massachusetts*, by David M. Belding, M. D., Massachusetts Marine Fisheries Series, Number 3, 1910.)

Perhaps because the range, season, and customary catch of the scallop is so limited, its natural history should be summarized briefly. Drawing again from Dr. Belding: length of a scallop's life, 20 to 26 months; legal size in Massachusetts: "must have a well defined annual growth ring"; spawns from June 15 to August 15; feeds on diatoms contained in the water; seldom found in more than 25 feet of water. Adding slightly to his facts: the legal period for taking scallops in Massachusetts is from October 1 to April 1, but since the management of the shellfisheries is left pretty much to the towns, the season may vary from town to town within these dates. The limit of the take is also controlled by the towns. (In Connecticut the dates are virtually the same, except in limited areas traditionally controlled by the towns involved.)

Bay or Cape Cod scallop, *Pecten irradians.*

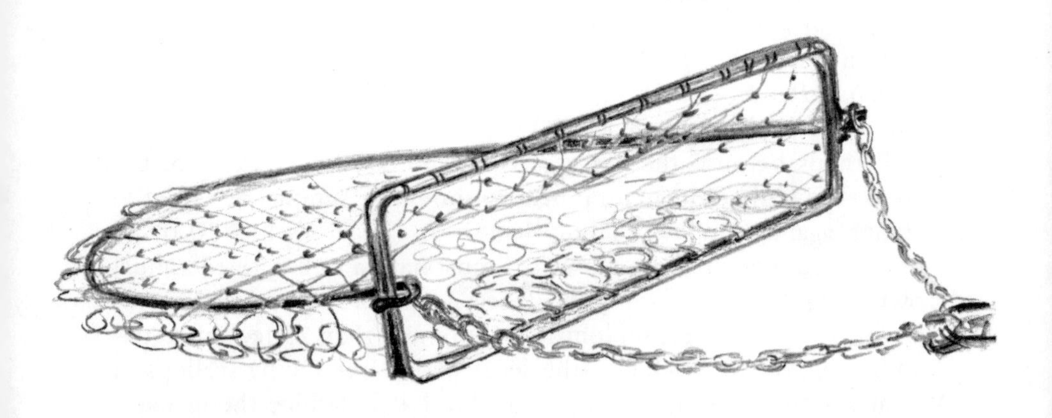

The original bar or box scallop drag. This is probably the first scallop drag. (It is now outlawed in Chatham, Mass., for no known reason.) It certainly dates back to the early 1900s. Fundamentally, the design is a component of the Nova Scotia or shenanigan, deep-water, sea-scallop drag.

Scallops may be taken by hand, and in deeper water by dredge (or, as it is more commonly called, by "drag"). Scallops in their younger stages are quite mobile, both crawling and swimming, but as the winter of their second year approaches, they are apt to lie in little hollows on the surface of the sand or mud, hollows that they create for themselves by flapping their shells.

The frame of a light scallop dredge or skiff drag has, with few exceptions, changed little since Dr. Belding's time. It is usually formed of a demi-hoop of ½-inch round iron, about three feet across. (This varies, depending on some town limits and on the individual fisherman.) To the rings or ears at the ends of this half hoop is attached a shorter loop of fairly heavy chain or a string of lead rollers. It is my own opinion (but it is agreed to, I think, by most experienced scallop fishermen) that the chain works better in sandy, shelly, rough bottom, whereas the rollers work better in mud and eelgrass. The distance from the center of the loop to a line drawn between the ears of the drag should be at least half the distance between the ears, no matter whether the loop is a chain or a roller drag.

Following this chain or line of rollers, which quite often is strung on light, flexible cable, is a chain net bag of two-inch rings made of 3/16- or ¼-inch steel. The rings are fastened together with softer iron connectors of the same diameter. They are oval, and about half as wide as they are long. They are sold open so they may be hammered closed, to connect two rings. The chain bag is usually made six to nine rings deep if the drag is to be brought aboard by hand, and it is exactly as wide as the drag frame when it is spread out.

Securing the chain bag to the leading chain, called "hanging the bag," is one of the fussier parts of construction. With the leading chain spread as it will be during fishing, the square chain bag should be pulled up so that it is straight, not curved, and just a connector link away at the center. There are then unfilled triangles at each side between the chain and the rollers. These triangles have to be filled with more rings and connectors so that the leading chain will not be pulled out of shape, and yet the chain bag must lie straight and flat. Starting from the center ring, which will probably be connected to the chain with a single connector, the next ring on each side will probably be connected with three connectors—not two, because this

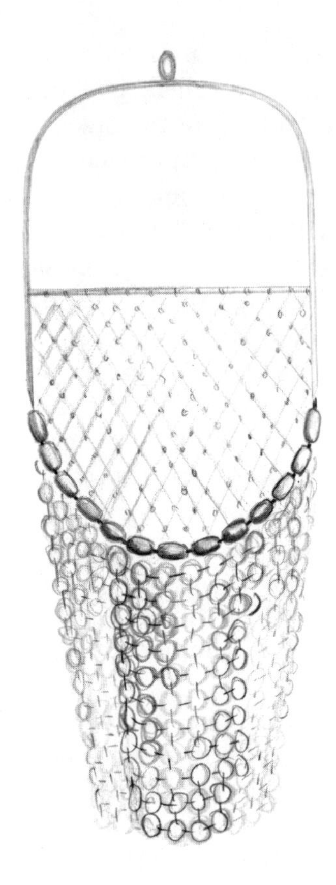

The basic scallop skiff drag, with lead rollers, viewed from the bottom.

Two-inch scallop-drag rings and connector. Connector is soft iron, so it can be beaten down to connect rings.

would tend to twist the connection. Next, a ring fills the space with a connector fore and aft, then perhaps two rings. The whole trick is to get the rings to lie flat and yet leave no unduly large holes.

While the whole rig is still laid out smoothly, a string back must be attached. A two-inch mesh of fairly heavy twine can be secured to the outer rings directly, of course, but the twine will last much longer if a length of ¼-inch nylon rope is first clipped to the rings with connectors and the twine is clove-hitched with a locking half hitch added at each connector all the way around. If you want to leave an ear, or a six-inch loop of rope, at each of the bottom corners, it will help you to dump the drag when it is brought aboard full of shells and trash.

Now you have a bag made of two-inch twine on top and two-inch steel rings on the bottom where the wear occurs. A reinforcing rod is frequently welded across the drag's demi-hoop, about a third of the way up, and the leading edge of the twine bag may be secured to the rod. Secure the drag warp to the leading edge of the demi-hoop, or to a ring welded to it for the purpose. Use up to ½-inch manila; the heavier the rope, the easier it will be on your hands when you haul back, but also, the heavier the rope, the more it will tend to float your drag as you tow it along, thus tending to lift it off the bottom. A ⅜-inch rope made of one of the new synthetics is a nice compromise. The rope must be strong: if you have a 16-foot boat with a motor pushing it on one end, and a drag full of shells and trash on the other, the rope must withstand the shock if you hang up on a rock, mooring, or whatever solid in the bottom.

As to scope, that is, length of the drag warp, all dragging is a nice balance between depth of water, length of rope, and speed. If you use too little rope or too much speed, your drag won't cling sufficiently to the bottom. With too much rope or too little speed, you'll fill the drag with mud and trash, and cover nowhere near enough bottom to bring home a paying load. Very, very roughly, I'd say, you'll probably want five or six times as much rope as the depth of the water. This varies greatly, of course, depending on the weight of the drag, the force of the tide run, and your speed, but it is a starting point. There is no formula regarding scope and speed; it's yet another trick that culls the men from the boys. As a start, suppose you have a 16-foot skiff with an 18-horsepower motor in 15 feet of water and are towing a three-foot drag that is nine rings deep and has bows made of ½-inch round iron. You'll probably want to have out 14 or 15 fathoms of ⅜-inch drag warp and will want to hold the throttle almost at the "start" position.

Try a five-minute tow and haul back. If the grounds you want to tow on are not big enough for a straightaway tow, circle back (not too sharp a turn, or you'll tip the drag bottom up). This circling will work to your advantage, because while the boat is traveling at the same speed, the drag will slow down on the turn. Haul back and see what you have caught. If the drag is full to the lead line, but the trash is clear of mud and sand, then you're probably about right. If the drag is full of mud, shorten the rope or the towing time, or speed up. If the drag is only half full, slow down, let out more rope, or tow for a longer period next time. This is for bottom that is relatively clear of grass.

If you haul back after a five-minute tow and the lead line is clogged with eelgrass and there's nothing much else in the drag, you have a problem that no one has ever really solved. You can try two or three things. The smartest and most productive solution is to try to start the tow in clear bottom before you get to the grass bed, and don't tow too long a time in the grass. One fairly common practice, which I don't like but which seems to work, is to tie a 10-pound grapnel to the drag warp about a fathom ahead of the drag. It tears up the grass somewhat, and it tends to make the drag jump. You can also tie a very heavy weight ahead of the drag, a couple of fathoms of one-inch chain or a half-dozen window weights. Theoretically, this holds the bow of the drag against the bottom so that the grass will catch on it instead of on the lead line. It's stupid dragging, but at times it will work when nothing else will.

There are four variations of drags used in my area. The first is the standard one that I have described; the second is very good, but it takes a great deal of skill to operate; the third is a bastard variation of the second; and the fourth I owe to my friend Bob Whiting, who invented it and is one of the very few to use it. It is good.

Some years back, Bob came up with the plan of putting a planing board on the forward end of his drag. The idea is to get a flow of water *over* the drag instead of through it, to create a sort of vacuum. Scallops, being relatively light, will lift to the vacuum, whereas the heavier stones and trash will not be dragged into the drag. All through one winter, we fiddled with planing boards. With

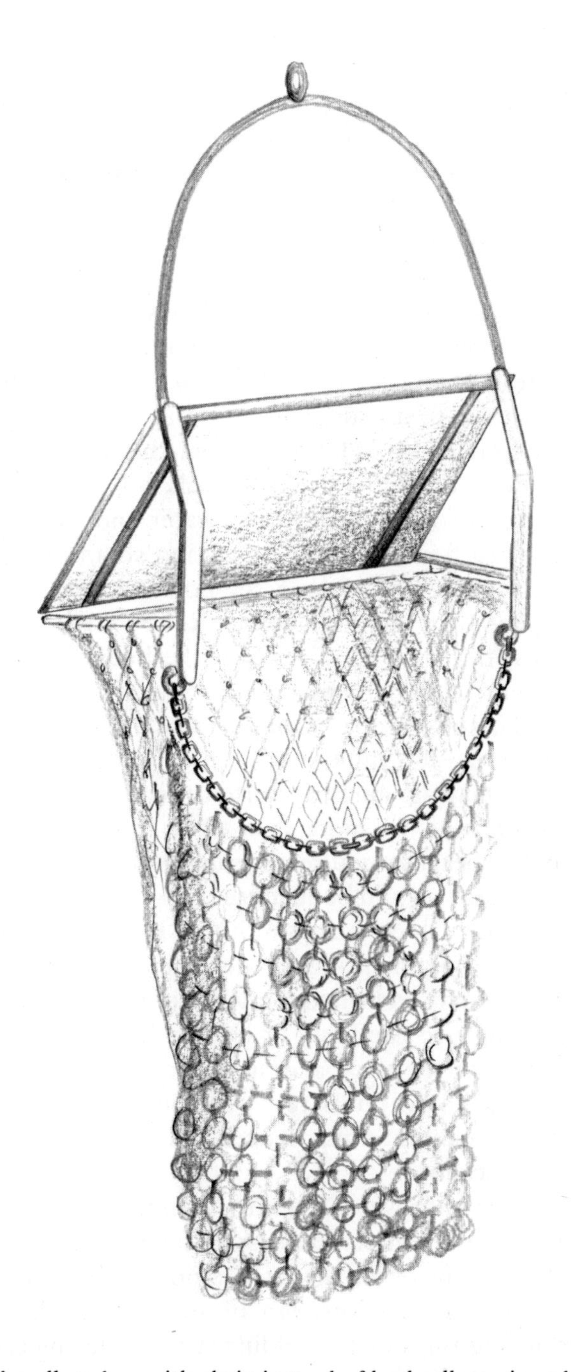

Planing board scallop drag with chain instead of lead rollers, viewed from below.

light, two- or three-foot drags, we arrived at a sort of formula. Later, in Cape Cod Bay, in much deeper water, we found that with a similarly light drag the same principle worked, though the drags, being 54 inches wide (the town limit in these waters) and towed with cable, needed considerable reinforcing.

The planing board may be made with an iron frame with plywood bolted to it, or with a ⅛-inch steel plate. First, however, one foot forward of the ears of the drag, the bows are bent upward so the leading, towing ring is about six inches above the level. A six-inch-high vertical frame is welded just forward of the ears to which the chain is shackled. A foot-wide piece of ⅛-inch-thick steel plate, or a piece of ¼-inch plywood, is fastened from this vertical frame to the bend of the bows. This is the planing board.

As the drag is towed through the water and along the bottom, the board tends to force the drag down and shoot the water over it. If the drag is towed slowly along the bottom, nothing particularly different happens than would happen without the board, but as the speed increases, the flow of water over the board increases, until it reaches the point where the vacuum is created and the scallops are sucked off the bottom. Scope becomes crucial; too little and the drag won't tend bottom, too much and you've defeated the purpose of the board. You may be able to shorten the scope so that the warp length-to-depth ratio is no more than three to one. The drag should fill more quickly because you're covering more ground in less time, and the shack (trash) should be lighter.

Unfortunately, a local blacksmith saw a planing-board drag and, without understanding the principle, decided to "improve" it. Having no 12-inch-by-⅛-inch steel plate, he welded a six-inch planing board at a 45-degree angle to the bow, or leading hoop. While a drag built that way works better than a drag without a planing board, this particular planing board tends to force the drag harder against the bottom, thereby making it dig deeper and catch more heavy shack but not as many scallops.

Bob's further improvement on the planing-board drag, in order to get around the extreme tenderness as far as scope and speed were concerned, was to make swivels at the point where the bows meet the planing board. The swivels can be made either by welding together interlocking rings or by welding two ears facing each other on each side, with a pin locking them together.

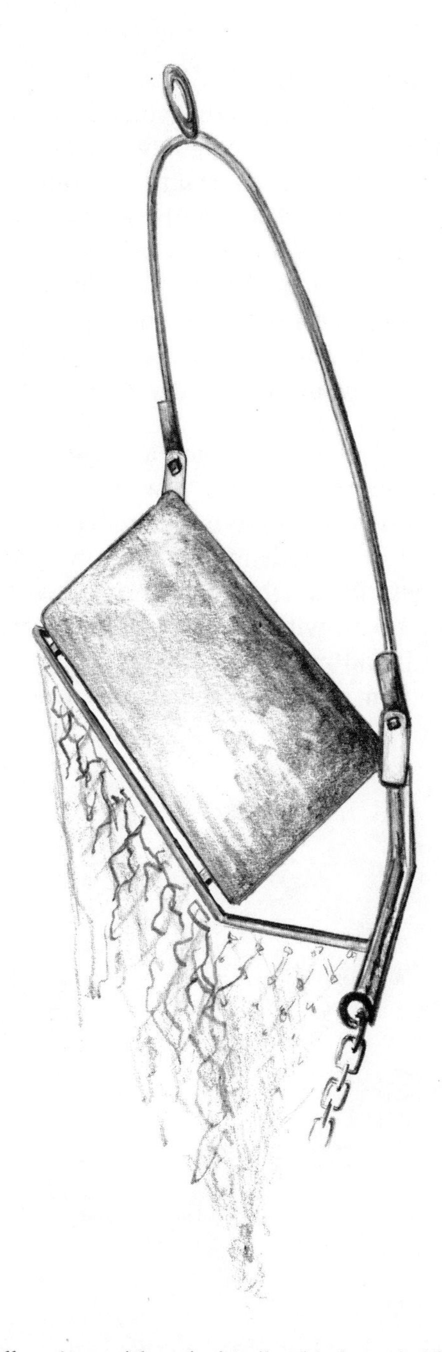

Planing board scallop drag with swivel bail. This is probably the most sophisticated drag to date.

It is a fairly common practice locally to rig a culling board athwartships on the skiff. The culling board is usually a piece of plywood four feet wide and as long as the boat is wide. This should be backed by two-by-fours braced against the sides of the skiff (because you may want to stand on the board to bring a loaded drag aboard). There should be a four- to six-inch railing secured at right angles to the plywood at each edge, to keep the water, scallops, and shack from falling into the boat. The culling board is built in, far enough forward to allow standing room by the outboard, but not so far forward that it will require too long a reach to the outboard tiller and throttle.

Most of the local draggers fishing a single drag haul over the starboard side. (In other towns, they tow two, four, and even more drags, the reason for which I have never understood, since fishing a single drag properly will catch more than any two men can keep up with.) So it is best to sheathe the rail of the skiff and the outer edge of the culling board with a piece of sheet metal of some sort. It saves the boat rail and makes the drag slide more easily onto the culling board. You can even rig a small pipe roller, the full width

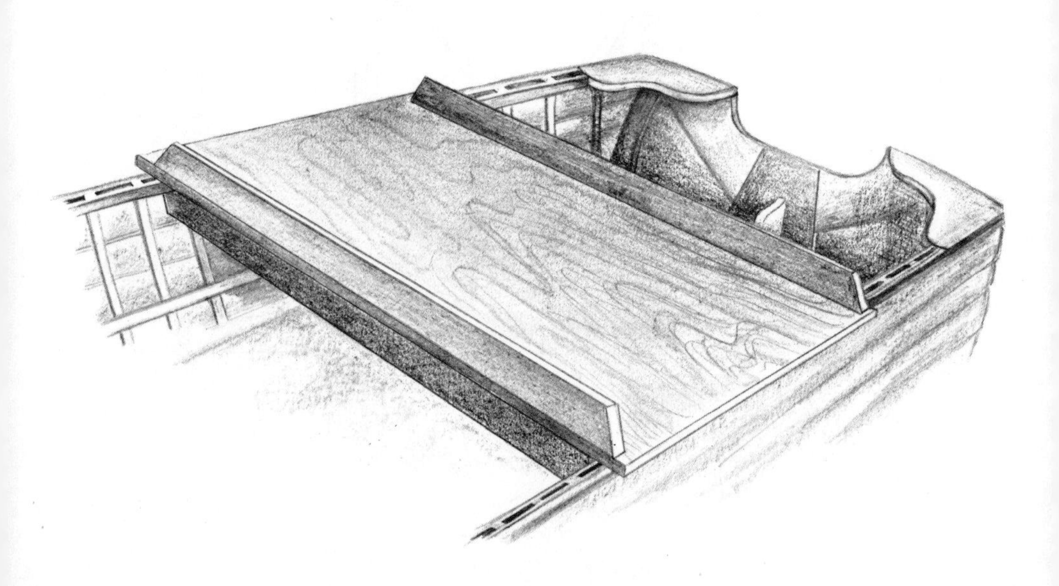

Culling board for scallops.

of the culling board, to make things easier, although few local fishermen bother.

When a tow is brought aboard and dumped on the culling board, the drag is thrown over, carefully right side up. The motor again is thrown into gear, the scallops are culled into a basket or bucket, and the trash is pushed overboard as you go. If the trash gets ahead of you, don't keep towing until the board is clear; haul back and hang the drag at the rail. (If you leave it on the bottom, using it for an anchor, scallops will tend to wash or swim out.) If you keep on dragging after the drag is full, you'll wash out the lighter scallops and fill up with heavier trash, and also stir up the bottom to make dragging harder on the next tow. Instead, fill the culling board, hang the drag alongside, and drift or jog off the grounds a little way to cull so that the shack doesn't pile up just where the scallops are densest.

As the catch mounts and your wire basket fills, wash the scallops before you dump them into bags. When the basket is half full of scallops, slosh it up and down over the side until it no longer streams mud and sand. This will make the scallop opening much easier and the opened scallops much cleaner.

To open any amount of scallops for resale, you must consult your local Food and Drug Administration for regulations and restrictions. (Scallops are sold shucked, whereas other shellfish are not.) Chances are you should have an inspected, certified, and approved shucking house, stainless-steel bench, pitched floor with drain, and so on, and so on. If you merely want to sell them to your friends and neighbors, however, chances are the restrictions are simpler. Nobody seems to care if you make only a few people sick; they just don't want you to do it wholesale.

You will need, in any case, a fit place for opening the scallops. If possible, you should be able to scrub down the area with a hose and a scrub brush when you're through. That's the easy way. Or you may be able to get away with opening the shells in your kitchen sink—and then repaint around the sink when the season is over. If you have a bench, it should be belt-buckle high, a little higher than the average sink and about 30 inches wide. The lighting should be good. You will need something for storing the opened scallops (which the fish buyer may furnish you) and a stainless opening pan that holds anywhere from two quarts to a

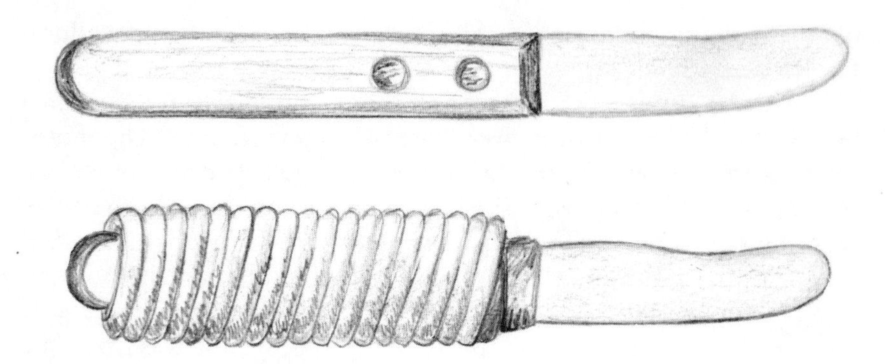

The shape of a standard scallop knife is good, but the handle is too small, so it can be built up with codline and black plastic tape, as shown.

gallon. And you must have a good knife if you're planning on opening any quantity. Round-ended scallop knives have only two faults: the blade is too short for big scallops, and the diameter of the handle is too small for most male hands. The shape is good and shouldn't be changed, unless you want to sharpen the cutting edge, but that's not essential.

I usually cut off the handle flush with the first rivet to lengthen the blade. My favorite knife has a handle bound with eight-pound codline to build it up (it also makes it slightly corrugated to keep it from slipping), and the whole handle is bound carefully with black plastic tape. It's smooth and it's clean, even if it's not "legal," and it is far cleaner than the common practice of driving the handle inside a four-inch length of garden hose.

You can wear rubber gloves to open scallops if you want to, but they'll slow you down and they're messy. I prefer to bandage my hands with surgical adhesive tape. If the tape is properly applied, even slightly heated over the stove or with a match when my hands are dry, it will last for four or five hours, or through four or five bags of scallops.

To tape up: On the right hand, take two or three turns of inch-wide adhesive just above the thumb joint; this is the pivot point. Next, take a couple of turns of inch-wide tape between the joint and the knuckle of the thumb, and the same between the knuckle and the second joint of the right forefinger; this is where

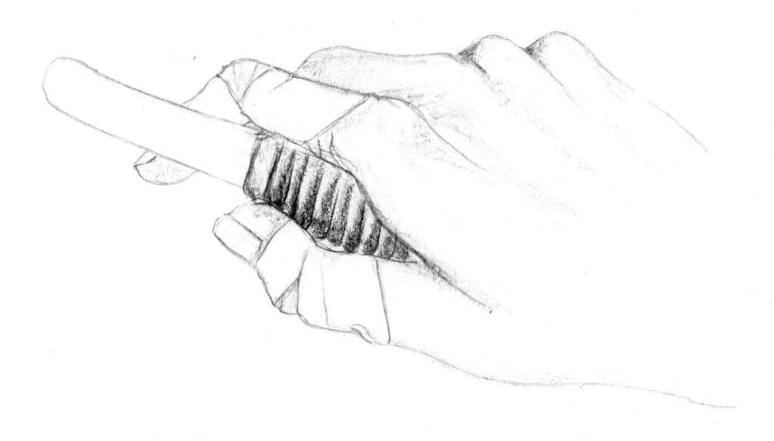

The right hand is relatively easy to tape for opening scallops.

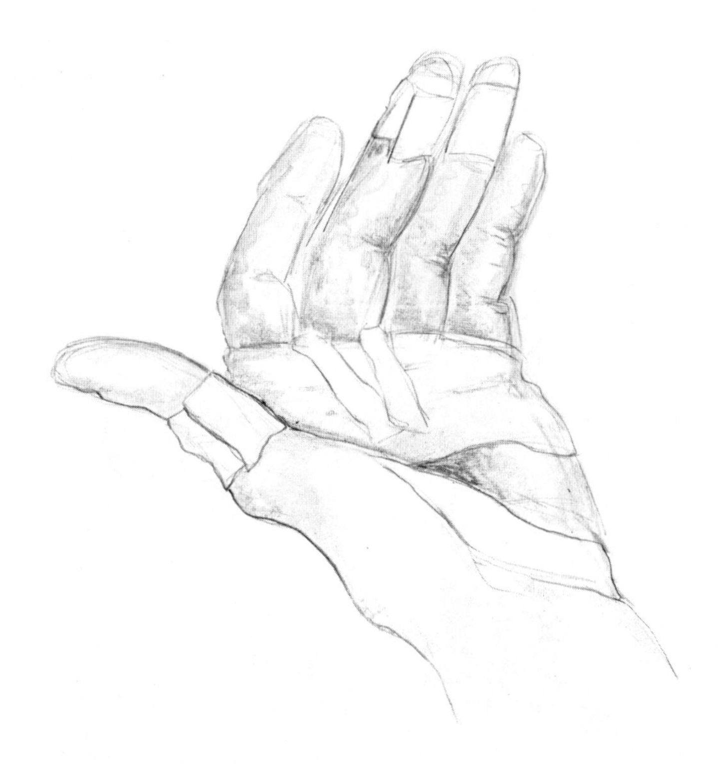

Protective taping for the left hand for opening scallops. Tape is placed at the points of greatest wear.

you'll be cut if the shell breaks or your knife slips. Then take a few turns of tape between the first and second joints of the forefinger. On the left hand, in the palm (because that is where the greatest wear occurs), put a couple of squares of two-inch surgical gauze, covered with at least a double thickness of two-inch adhesive. Don't put it on too tight—just enough so that the edge extends above the junction of the fingers and the palm. Between the first and second fingers, and between the second and third, tape a ½-inch-wide strip long enough to lock the two-inch-wide piece in place. Then put an inch-wide piece of tape just at the first joints of the second and the third fingers. Then, as a precaution, put a strip of two-inch tape across the fleshy area just below the left thumb, anchoring the strip on the thumb, between the joint and the knuckle. This seldom stays in place, but after you've slipped once or twice and carved yourself up, you'll be more careful.

To open scallops: Hold the knife in your right hand, with the cutting edge facing in. Pick up a scallop with your left hand and hold the hinge away from you, dirty side up. You will notice that

There are three steps for opening a scallop shell and removing its contents.

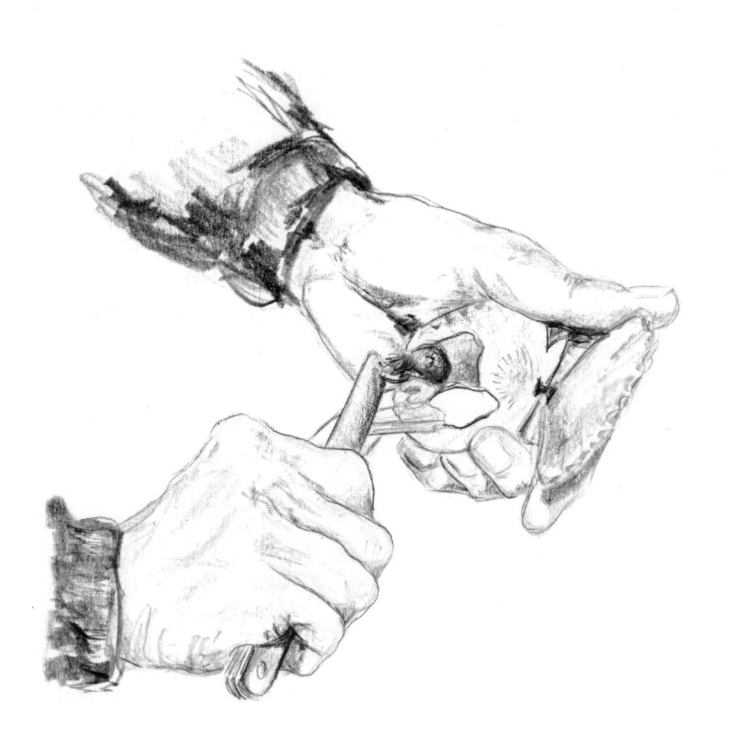

the edge of the shell on the right of the hinge is straight, while the edge on the left is indented. If the scallop is equally clean on both sides, as sometimes happens with scallops living in clear sand, make sure the straight edge of the scallop is to your right, the indented side to your left.

Put the padded ball of your right thumb on the top shell, holding the knife between the tips of your fingers and the palm of the same hand. Bring the knife point to the opening between the shells, just where the straight edge swells to the shell proper. Insert the point and slightly twist the edge up and the back down. If the scallop shells are brittle, this twist should be very light, or it will break a corner off the top shell. By twisting the knife slightly and sliding the point pressed against the top shell halfway through the scallop, you should be able to cut the adductor muscle clear of the top shell. A slight flipping motion away from you should lift the top shell clear. Now slide the knife point in against the bottom shell, under the guts, at about "two o'clock" (reading the scallop as a clock, with 12 o'clock at the top, away from you). Slide the knife point counterclockwise around the adductor muscle, from two to 10 o'clock. As the point reaches 10 o'clock, lift and flip the gut toward you, squeezing the edge of the gut lightly between your right thumb and the knife blade. If the scallops are fresh and you do it correctly, the gut will come clean. Then scrape the adductor muscle off the shell with the knife blade, scraping toward you and using your right thumb as a kind of stop. A flick of your right wrist will pop the scallop—or the part of it you want—into the stainless pan. Before it lands, you should have dropped shell and guts into the barrel between your legs, and your left hand should be reaching for another scallop. With good scallops (no more than 140 to the pound), a good opener should be able to shuck slightly more than a gallon (nine pounds) an hour.

Do not wash the scallops in fresh water, no matter how dirty they may be. It's against Massachusetts law in the first place, since they absorb a great deal of the water (as much as 30 percent of their weight). Also, the water makes them swell and spatter when they are fried. Slosh them in their own juice if you want to make them look better. If it's very early in the season and the scallops are very muddy, go to the shore and get some clean salt water and rinse them lightly. Store the scallops in a cold place above the freezing level and get rid of them as soon as possible; they're never as good after the first day.

If you want to freeze them for your own use, it is best to quick-freeze them in their own juice. If you don't have a quick-freeze unit, spread them out one scallop deep in an ice-cube tray, freeze them, and then wrap them in freezer paper, flat. With this method, thawing is quick and packing is handier.

So far, I have dealt only with scallops caught in a drag or dredge towed behind an outboard-powered boat. There are times and places in wadable water where picking up scallops by hand is easier and quicker. This is particularly so where water is no more than knee deep at low tide and the bottom is covered with thick eelgrass.

My favorite way of scalloping under these conditions is to get out of my light skiff upwind of the grounds to be fished. With the skiff drifting broadside behind me to smooth out the waves so I can see the bottom and the scallops more readily, I wade along before the wind, dip-netting the scallops as I come to them, dropping them into the skiff as they collect in the net. I use an ordinary, cheap, herring dip net, which is fairly strong, light, and easily maneuverable. This works best if there are not too many people around to roil up the bottom, and also if you can drift, so to speak, against the tide, so your own roil is always behind you.

One gadget that seldom is used in these parts, but is highly popular in other places, is a "looking box." This is simply a

Scallop looking boxes.

wooden box with a clear glass bottom and no top. It is my understanding that most of the looking boxes are built around a pane of glass, 10 inches by 12 inches, or 12 inches by 14 inches—whatever size pane is handiest. The box should be watertight, and the easiest way to make it seems to be to rabbet out the wooden sides, line this groove with a seam compound, and then fit in the glass. Some looking boxes are quite complicated. Some are carried by a strap around the fisherman's shoulders, thus leaving both his hands free. Some even have a cloth shield or funnel to keep the sun from reflecting into the fisherman's eyes.

There is one other tool that is most handy but has almost never been used locally in recent years—perhaps because there were few scallops in shallow water for so long after the eelgrass disappeared. It is a scallop pushnet or scoop, simply a demi-hoop made of iron a

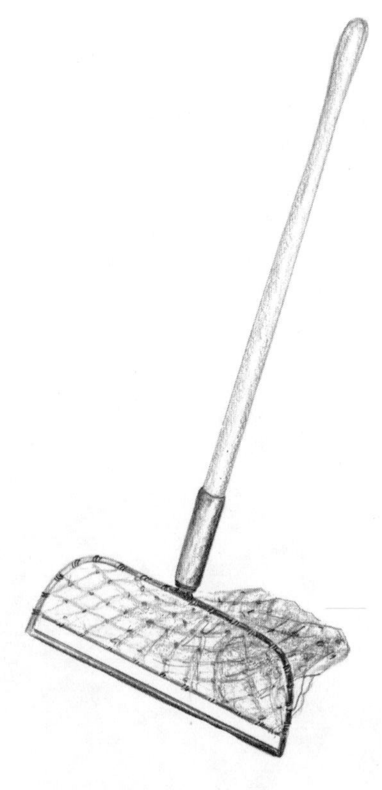

Scallop push-net or scoop.

half-inch or so in diameter, secured by a tang to the bottom end of a 10-foot pole. (The length is not so important, except that the pole should be long enough and strong enough to give the fisherman a good grip and leverage.)

Across the leading edge of this 18-inch- to two-foot-wide hoop is a straight bar. It can be made of ½-inch half-round iron, or it can be a flat piece, perhaps an inch wide and ⅜ inch thick. In the former case, you may want a parallel bar welded about ½ inch back; in the latter case, the flat bar should be perforated on the back edge at two-inch intervals. It is to this parallel bar or to the perforations that a net bag is hung, looping back a foot or so and secured to the upper bow or demi-hoop. The net should have a mesh small enough so that the scallops will not wash through. You have, then, in effect, a dip net that is more solidly built than most, and one with a straight leading edge instead of a round one.

This net is simply pushed through the eelgrass, so it digs into the mud as little as possible but scoops up all the scallops in its way. I don't know why more modern fishermen don't use this tool. Perhaps it is because it is one of the old-time pieces of equipment that was forgotten when mechanization took over.

Scallops taken this way, of course, should be handled just like any other scallops—wash them thoroughly at the shore to save yourself trouble when you get them home to open them.

7 BLUE MUSSELS

Every once in a while there comes from a sort of limbo a new way of making a living. In spite of the fact that blue mussels have been commercially profitable in Europe for centuries, there has never been a large demand for them in this country until recently. Even as I write this, some towns in Massachusetts are paying to have mussels destroyed, while other fishermen have seen the light and are making a good profit, especially in the winter, when most shellfishing is in a slump.

Blue mussels (*Mytilus edulis*) may grow on almost any kind of bottom, from coarse sand to silty mud. They are very tolerant of tide flow, from comparatively fast tide run to almost stagnant water. They grow on flats that are exposed to the air more than half the time, and also in water sometimes as deep as 10 fathoms. They will attach themselves to lobster pot warps, to docks and spiling, to rocks, and even to mats of vegetation hardly strong enough to hold themselves together without the help of the mussels. Being very much a colony shellfish, mussels will grow on top of mussels until they smother each other.

In former years, mussels brought the energetic fisherman as little as 50 or 60 cents a bushel, but the popularity of these shellfish has increased so much that there are times when they bring more than adult quahogs in the wholesale fish markets.

The tools needed to gather blue mussels are simple and inexpensive, and they are often the ones used to collect other shellfish.

A blue mussel, *Mytilus edulis.*

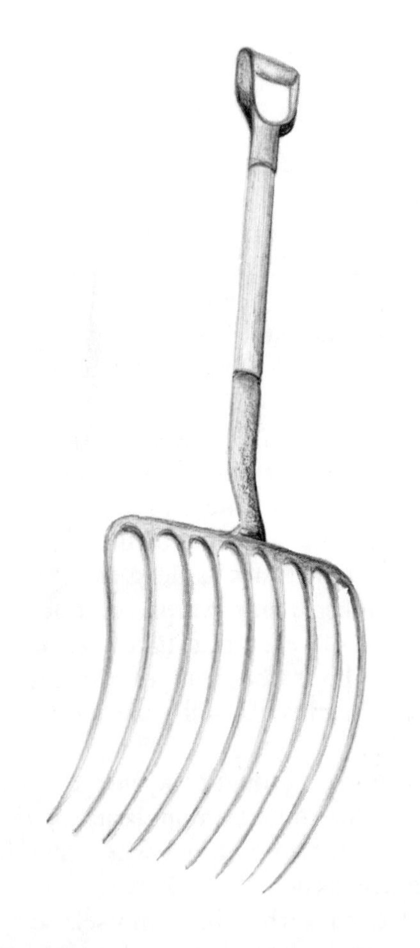

A coke fork, the most efficient tool for taking up blue mussels above low tide.

Depending on the area in which they grow, they may be picked up by hand, scratched up with a quahog scratcher, scooped up with a coke fork, tonged or long-raked, or dragged up with scallop gear. Since they grow in clusters on the surface of the bottom, a simple cruise through an area at low tide will reveal what mussel beds are available.

Aside from having a good, rugged skiff and the inevitable shellfish license, plus the tools adapted to the depth of water in which you intend to work, all you will need to do is establish your market. In the summer this may take a little doing. Take a sample—a well-washed, culled sample—to the local fish market. Inquire at the fisherman's cooperative or the fish buyer's. Write or phone a commission man in the Boston or New York markets, guaranteeing a supply. There seems to be more demand for mussels in the winter, but this preference is rapidly giving way to a year-round market.

Wash the mussels well before trying to market them. They are attached to each other by "byssi," or beards, and these strong hairs collect mud, which will detract from the market value. If the mussels have attached themselves to too much trash, dead shells, stones, and the like, you will do well to cull out the worst of it. Do not break up the clusters unless the local market insists on it. The byssi extend into the mussels themselves, and if you try to separate the mussels, you can pull out the threads from inside the mussels and thereby kill them, or at least damage them so that their life span will be scarcely long enough for you to market them alive.

Most towns specify no minimum size on mussels, but most markets do. Don't try to market mussels that are less than two inches long—2½ to three inches is a better size. Extra-large mussels may have a market, but most aficionados find them tough and strong. They do make good trawl bait, however, when trash fish such as skates and dogfish are plentiful, so you may be able to find a profitable market for the extra-large mussels with the line trawlers at certain times of the year.

Perhaps one of the reasons for a prejudice against blue mussels is that they are among the first filter feeders to ingest *Gonyaulax tamarensis,* the dinoflagellate that causes the so-called red tide in northern waters. If you plan to do extensive musseling, check with the local Department of Public Health or the Department of Environmental Resources. They will make a spot check for you (for free!) to make sure you will not be responsible for making someone sick. (You may have to furnish the mussel samples, however.)

Thus far I have made no mention of mariculture, shellfish grants, or raising your own shellfish for market. This is probably the best place to deal with the subject, since mussels seem to be the easiest and most productive crop, if not the most expensive, to raise in a saltwater garden plot.

The grant laws in each state in New England vary; for the most part, the grants are restricted to residents of that particular state or a corporation based in the state. The Division of Marine Fisheries or a comparable authority will be glad to tell you the regulations. It is best to check with them before you start.

Now, assuming that you have cut the red tape and are the lessee of a private grant, you can raise your mussels in three ways. First, catch your own seed on ropes suspended from rafts in anywhere from six to 40 feet of water. Second, if seed is available, you can tray it in trays suspended from rafts, no matter what the depth of water, provided the trays don't rest on the bottom at low tide. And third, you can simply transplant juvenile mussels from a flat that is exposed more than half of the time to an area with good growing potential (i.e., plenty of food, a reasonable tide flow, relative protection from ice in the winter).

It is well-nigh impossible to give concrete figures on the cost of rafting mussels. To be practical, each raft should be able to hold at least three trays. A three-tray raft requires: styrofoam logs for flotation, lumber for the trays, wire screening, strapping for reinforcement, staples, nails, anchors, and rope. Altogether, this rig may run just over $200.

Anchor the rafts fore and aft over your grant and forget them until the market is ready. If your water is as productive as ours, mussels that are an inch or so long in April or May will double in size before winter.

If each tray holds five bushels of seed, that much should produce double the amount of salable stock. If the mussels sell for $5 a bushel wholesale, each tray will pay for itself with no profit the first year. But after the first year, it will return your original investment, and from then on for the next five years, there will be no cost except your own labor, which is minimal. You need only gather the seed and market the prime, clean stock. The system has its advantages: the mortality should be negligible, because the predators can't reach your crop, and there is no trash to detract from the quality of your product. When the mussels have grown to the desired size, you will have no difficulty in holding them or harvesting them once the price is right.

A word about very small mussel seed: it won't stay put. If you put it in a tray roped tight to the styrofoam logs, after a month you are apt to find it has crawled out of the trays and attached itself to the styrofoam.

If your grant is in relatively shallow water, six feet or shallower, but deep enough to float the rafted trays at all times, and if the bottom is solid, not gooey mud or rapidly shifting sand, there is no reason not to cover the bottom with juvenile mussels, holding them until they are big enough for market. They will probably not grow as fast as the rafted stock, and, of course, they will not be as clean (since they are subject to predation, which the rafted stock is not), but they will be where you can put your hands on them when the price is up.

There are other traditional ways of growing mussels, such as on heavy poles pumped or driven into the bottom, or on ropes suspended from rafts. Both methods might grow more mussels per acre, but both are dependent on getting a natural set of seed mussels on the structures, which is not as certain as gathering your own stock and spreading it uniformly.

8 PROFITABLE PREDATORS AND ODD WAYS

There is hardly anything along the shore in which and for which there is not a profit and a market: moonsnails; whelks; and all the crabs—horseshoe, green, and hermit (the last is the only one, to my knowledge, that is not a predator). While they all inhabit tidal estuaries and sheltered water, their proliferation will depend on the particular locality.

MOONSNAILS

Moonsnails are called "round wrinkles" locally, and they are most common hereabouts. They are the worst predators and the most profitable to gather. If there is an appreciable population, some fish buyers will find a market for moonsnails for food, and you can unload them there. The large ones, 1½ inches or more in diameter, are quite commonly used as codfish bait by the handliners at certain seasons of the year. They may be caught in pots or picked up by hand along the edges of the flats and on the flats themselves. We always found them best if we could get an early low tide, only slightly after sunrise, particularly on foggy, muggy mornings.

If you don't find them wrapped around juvenile clams on the

A moonsnail (called a round wrinkle on Cape Cod) is valuable as codfish bait in the spring.

surface, look for a short, wandering trail ending in a hump just below the surface of the sand. Any moonsnail that is too big to fit between the teeth of a standard quahog scratcher is big enough for market. You should be able to convince the local shellfish constable that he should pay a bounty for small ones. That, of course, will limit your catch the following year, but it will also save a far more valuable crop of clams for next year.

We augmented our catch by building pots that measured roughly two feet long by a foot wide by 10 inches high. The pots were made of used laths spaced no more than the thickness of the laths apart. Laths seemed to make better moonsnail pots than pots made of cellar window wire, since they gave the predators a better chance for suction as they crawled up the sides (we used the same pots for conks in another area). We slanted the sides slightly inward, so that the ends, a foot wide at the bottom, were no more than eight inches wide at the top. We lathed over part of the top, leaving an opening the length of the pot and about five inches wide in the middle. The pots were ballasted with four red bricks, or a strip of cement an inch or so deep and four inches wide poured into the pot on each side. The pot was buoyed with six-thread rope and a small float or net cork. We baited with chopped-up horseshoe crab or almost any fish gurry (fish waste). The pots need not be tended every day, but, as with any potting, the more often you tend your traps, the more fish you'll catch.

Moonsnails can be kept in crates or cars in salt water until the market is right, but don't make the mistake of trying to keep them where there is too much change in salinity or water temperature.

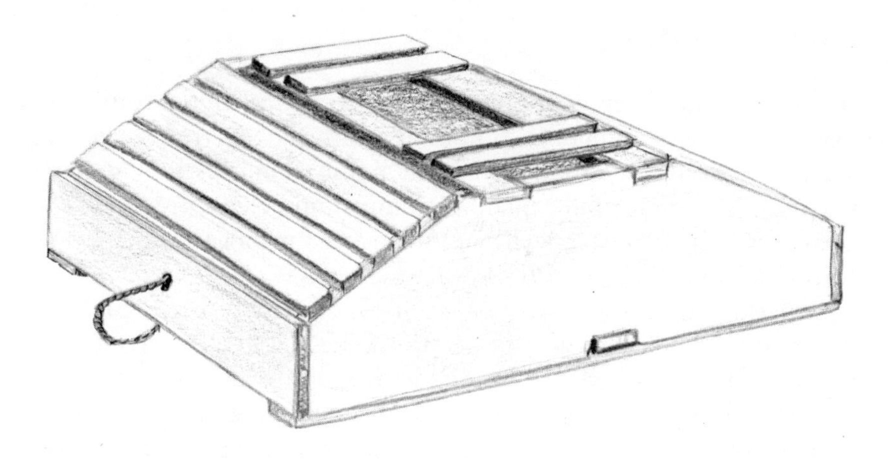

Whelk or conk trap designed by J. C. Hammond, Chatham, Mass., oyster grower. The trap can also be used to catch moonsnails.

CONKS OR WHELKS

These shellfish may be picked up or potted in the same way as moonsnails. Consult your local fish buyer and/or your local codfish handliners to find out which season is the most profitable.

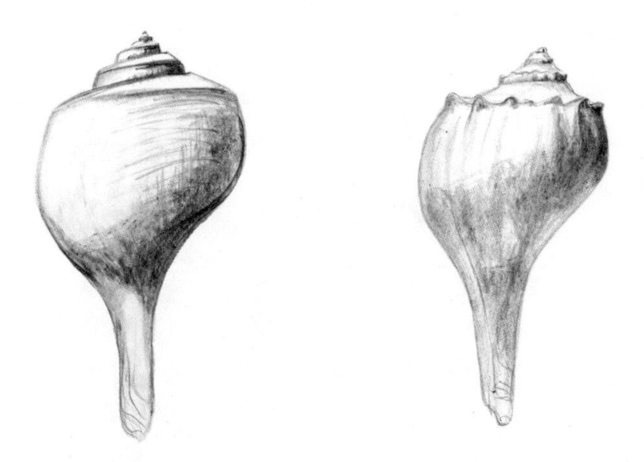

Channeled whelk, the soft-shelled variety (left), and knobbed whelk, the hard-shelled variety (right), make good codfish bait.

GREEN CRABS

Apparently the green-crab population is a cyclical thing; some areas may be infested, while an adjoining area may be relatively free. They can be deadly to the shellfish population, destroying clams, mussels, and even adult quahogs (almost unbelievably, they chip away the edges of the shells of full-grown quahogs until they get to the meat).

Mashed-up green crabs can be used for bait for eels, conks, or moonsnails, but there is a more profitable market for green crabs for bait wherever there are headboats fishing for tautog.

The traps vary, depending on the fisherman and the locality. I have a green-crab trap made of one-inch galvanized wire. It is two feet long by 15 inches wide by 12 inches high, with a six-inch-square opening in the top. The wire is turned down five inches all the way around inside the opening. I should think this could be varied, depending on tide and bottom conditions and the availability of materials.

To catch green crabs, use any fish or shellfish for bait, including horseshoe crabs. Recent research has indicated that maybe—just possibly—there might be a correlation between warming trends in the water and the quantity of the green-crab catch. I do not know that this is so, but I offer it as an element to watch.

The green crab is a pest, but it makes good tautog bait.

HORSESHOE CRABS

The continuing argument between those scientists who contend that horseshoe crabs are an endangered species and the shellfish constables concerned with the destruction of juvenile soft-shell clams by "horsefeet" is not going to be settled here or in the classroom. While horsefeet may seek shallow water in which to lay their eggs, the fact remains that they are commonly dragged up by quahog draggers and even sea scallopers in five to 10 fathoms, and this alone seems to indicate that the danger of their extinction is minimal. Whichever side of the argument you take, horseshoe crabs are without question the best bait for eels in certain areas, and therefore they have a market—sometimes as high as 25 cents apiece for large females. Chopped up, they are also good bait for conks and moonsnails. While I have heard talk of horseshoe-crab weirs, the crabs usually are picked up by hand locally. They can be stored almost indefinitely in a leaky, half-sunken skiff or dory.

CHUBS

You have a bunch of eelpots. Why not put them to work in the winter when the eels have stopped running? You can either leave in the center funnels or take them out. Tramp along the edge of the salt marsh some snowy day. You will find that there are certain slough holes and mosquito-control ditches that have a steady stream of water running through them. Find a hole deep enough to cover an eelpot at low tide. Bait the pot with a few broken quahogs, a gob of fish gurry, or even some leftover heels of bread. If you've found the right ditch, you will be amazed at the number of salt- and brackish-water chubs that you can pick up night after night. The freshwater, through-the-ice fishermen will gobble them up for bait, although you may have to shop around a bit to find a market. Try the bait stores closest to the cities, or the bait stores near popular freshwater ponds. The one requirement for holding your market is reliability. No bait-store owner is going to put up with you if you fail to show up when his customers are expecting bait.

There may be better ways to keep chubs, but the best rig we

found was a series of matching two-inch-high trays, about two feet square, bottomed with burlap (bags that the stitching had rotted out of during the quahog season). Taking care that they weren't touched by frost, we spread the minnows one layer deep, covered them with a thin layer of rockweed, and then hurried them to the tackle and bait shops.

Some minnows stand this treatment better than others. There is a smooth-sided minnow found chiefly in brackish water that is the toughest of the lot. Those with vertical stripes will stand little abuse, and likewise those with strong horizontal markings. I have no idea of their scientific names, but you'll find that some species survive a lot longer than others.

While we're on the subject of minnows for live bait, there is always a market for them during the summer months, either for freshwater bait, or—if there are fluke around—for saltwater bait. If you have children who are shore-wise, buy them a good 30-foot minnow seine. Notify the local tackle store that the kids have minnows for sale, and if they work at it, they'll be as financially independent as any youngsters in the neighborhood.

(We were fortunate enough to live on the shore of a fresh-water pond and found that most of the minnows would stand the transition. The kids stored them in some of my unused eel cars, which were made of cellar window wire. Of course, if you are properly located, you can hold them in salt water and not even have the loss we did in transferring them.)

At the same time they are seining minnows, the children may run across a guzzle or channel that is loaded with grass shrimp. This is a money crop. My kids, seining minnows for market, seined shrimp only on order, feeling that the supply was limited. Their customers keep coming back, years later, highly indignant that the kids have grown up, moved away, gotten married, and have kids of their own.

FIDDLER CRABS

Fiddler crabs (soldier crabs, army crabs, *Uca pugnax*) are the little, one-inch fellows with the equally large claw held threateningly over them. They range in color from a kind of enameled blue and pink to a dull brown. (One interesting sidelight: if a fiddler loses his large claw, which is usually on his right, he can grow another, but this time it is his left claw.) I say "his" intentionally, because the bait man we eventually did business with would take no females for bait. "If you catch all the mamas," said he, "there won't be no babies next year." That was the practical kind of conservation that the old-timers understood. The "mamas" have no large claw.

Fiddlers live in large colonies in the marsh. Since they are vegetarians, we found no way to trap them, though we dug all kinds of pitfalls and hedges.

We never had a very active market locally for fiddler crabs, but when an old bait man traveled all the way from Providence, Rhode Island, to Cape Cod for fiddler crabs, we got into the act. There is no need for equipment except a smooth-sided pail that holds a couple of gallons or more. We discarded our boots and waded into the marsh, wearing sneakers and jeans. After one or two muggy days, we decided a cotton shirt provided enough protection from sand flies and midges to compensate for the sweat and overheating.

Our bait man was so anxious to get fiddlers that he drove 100 miles each way, twice a week. This meant we had to hold our daily catch. Again we pressed our eel cars into use, trying not to crowd the fiddlers. Held this way, the crabs can survive underwater during a normal tide, but we lost six gallons of crabs on one set of high-course tides because they were underwater too long and drowned.

The fiddler (or army or soldier) crab is valuable in most areas for tautog bait.

I shudder now to remember the way the old man culled the live crabs from the dead: he simply plunged both hands in and let the crabs pinch onto his fingers. He figured that those that didn't bite him were dead.

The old man carried them the way we carried minnows, except that his trays had down-curving collars an inch or so wide.

You will find fairly easy picking the first time you go marsh trotting. Not expecting any trouble, the crabs will hold still, waving that big claw threateningly. But the second day, they will be a little warier, and by the time you've bothered them for a week, the survivors will pop into their holes the minute your head sticks over the edge of their slough hole.

HERMIT CRABS

If you live in an area where there is any amount of tautog fishing (blackfish, white chin, *Tautoga onitis*), there is always a market for hermit crabs, the big-clawed, soft-bodied crabs that live in other creatures' shells. The easiest way to pick them up is to wade off at low tide with a wire basket or a burlap bag, snatching them off the sand with a quahog scratcher or with your fingers.

The hermit crab is the very best tautog bait in Cape Cod waters. It is used as sheepshead bait in Florida.

The old-timers made a hermit-crab trap with the demountable rim of an auto tire for a frame, covering the bottom with cellar window wire or fine netting and crowning it six inches or so high to a four-inch-square opening at the top. Across this opening was strung a single wire for impaling bait: a half-dozen crushed hermit crabs, a few broken red paddle crabs, a chunk of horsefoot, or a fish head.

If you want to car the crabs, one of the eel cars will do, but be sure to include in your car a small tautog or almost any other live saltwater fish. A big eel will do, if there is nothing handier. This live fish, moving about in the car, will keep the hermit crabs in their shells and therefore alive, whereas, left to themselves, they will crawl out of their shells and die.

ODD WAYS

The many odd ways to whack out a living alongshore depend on your willingness to go your own way and do your own thing by yourself—not necessarily always for a full-time living, but as a way to make a buck on the side.

9 A BOAT FOR 'LONGSHORE FISHING

Probably the most expensive, and certainly the most important, piece of equipment you'll need to make a living alongshore is a boat. There has never been a boat that is perfect for all kinds of use, so you'll have to compromise. For all the lighter kinds of fishing described in this book, a 14-foot, flat-bottomed wooden skiff is best. It is adequate for everything except bullraking, flounder dragging, and scalloping. (While I have dragged scallops and bullraked from a 14-footer, a 16-footer, with its increased beam and higher sides and stern, is far better.)

Having used aluminum, fiberglass, and plywood, I still prefer a cross-planked wooden boat. However, in the search for material for this type of boat, I found cedar almost unattainable, pine of too poor quality to consider for boatbuilding, and the only alternative was mahogany. The plans for the 14-footer shown here call for plywood, and while a boat built of plywood tends to "scale" because of its relative lightness, plywood seems to be the logical material to use. The cost for materials for a 14-footer should be somewhere under $125.

Although the plans are specifically for a 14-footer, a 16-footer may be built by the same process, with the dimensions expanded proportionately. The boat is slightly "rocker-bottomed," since I

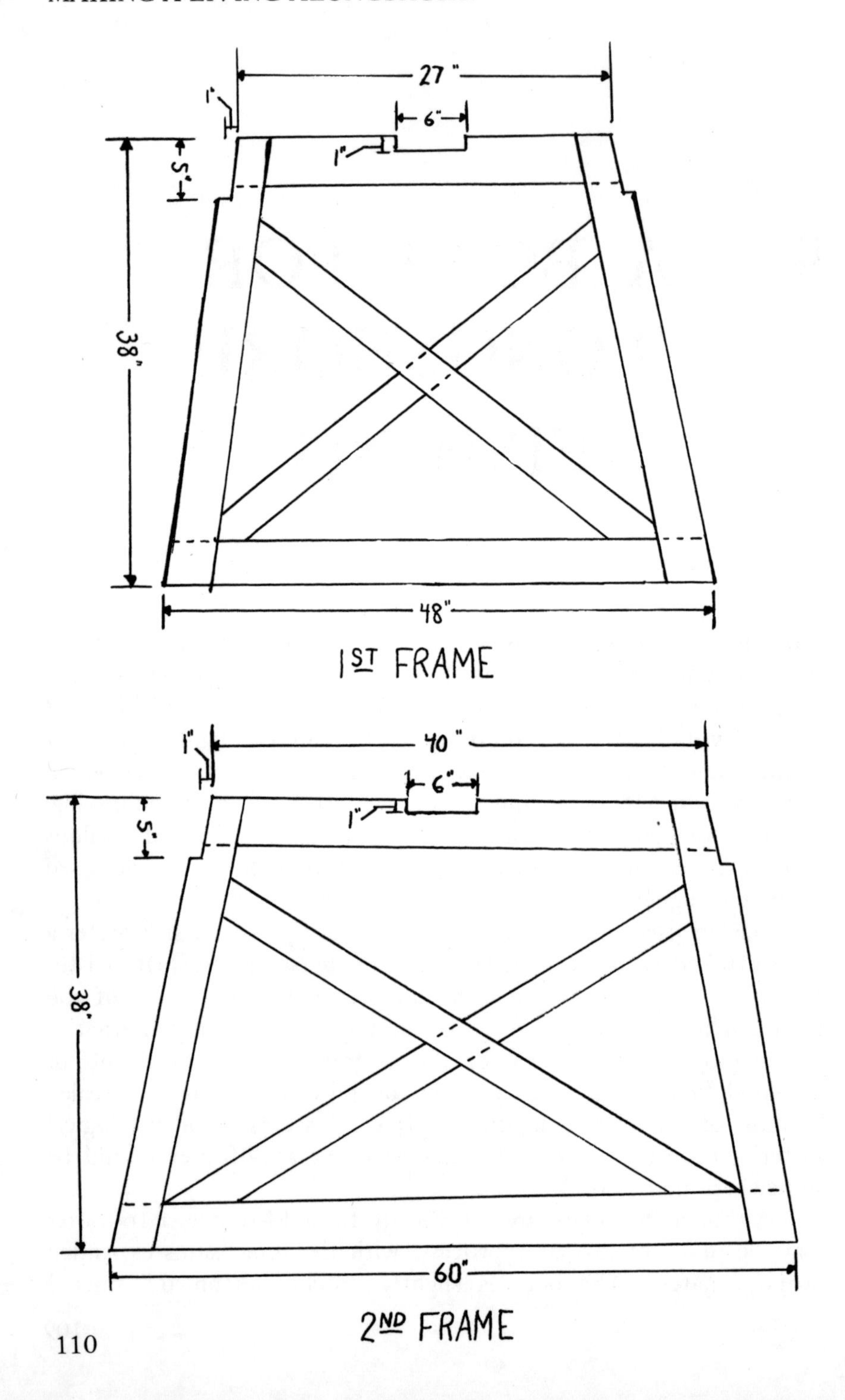
27"
1"
6"
1"
5"
38"
48"
1ST FRAME
40"
1"
6"
1"
5"
38"
60"
2ND FRAME

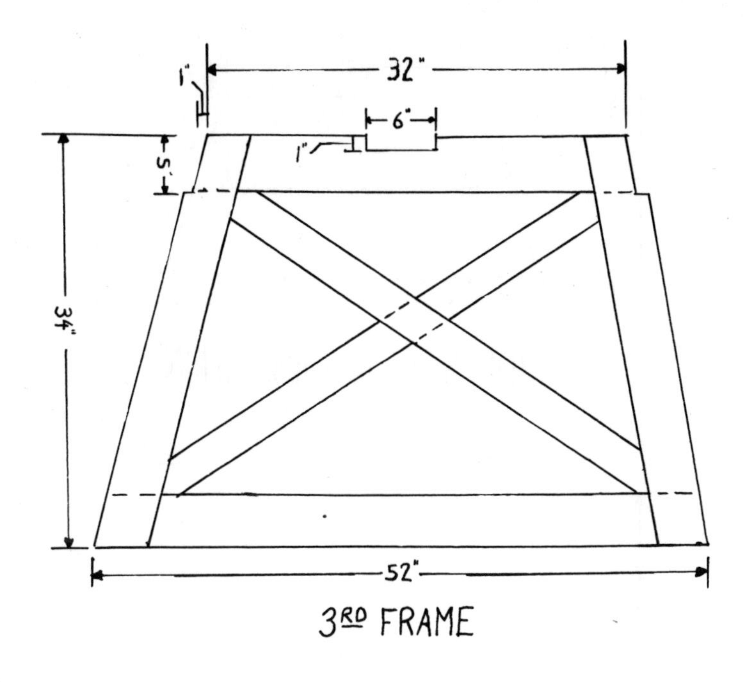

3RD FRAME

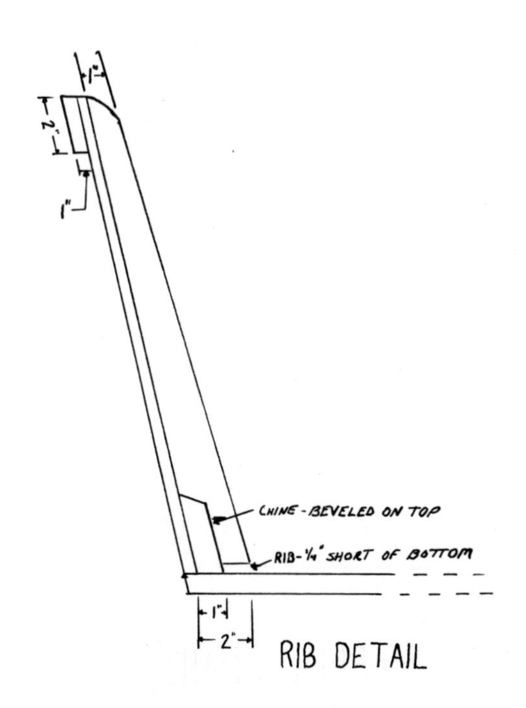

RIB DETAIL

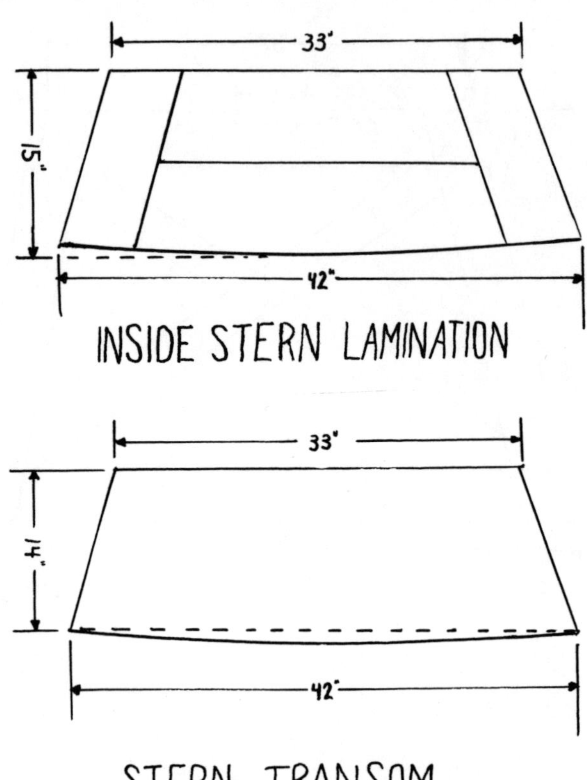
33"
15"
42"
INSIDE STERN LAMINATION
33"
14"
42"
STERN TRANSOM

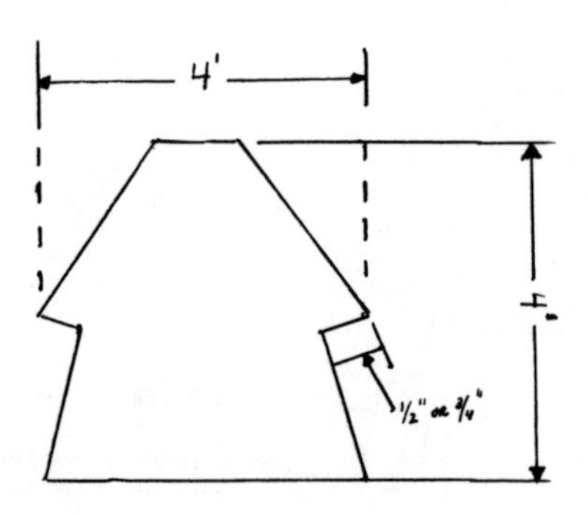
4'
4'
1/2" or 3/4"

STEM DETAIL

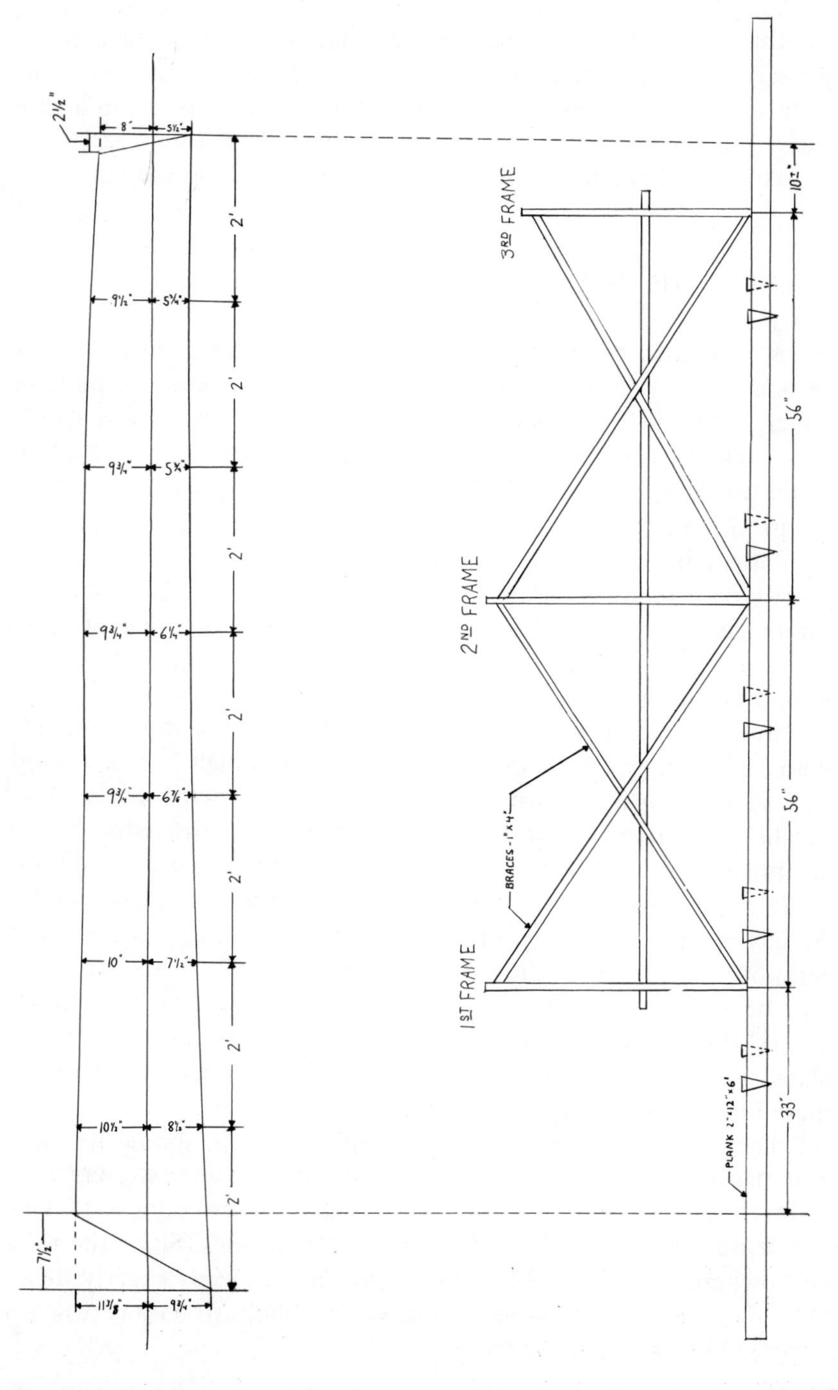
3RD FRAME
2ND FRAME
1ST FRAME
BRACES - 1" x 4"
PLANK 2" x 12" x 6'
56"
56"
33"

find this design most practical in very shallow water. In fact, it can be dragged behind you quite easily in wet mud with almost no water at all. The plans for the 14-footer have evolved from a skiff that reputedly was quite common in the Chesapeake Bay area some 30 years ago. They may still be used there, for all I know.

BUILDING THE BOAT

Unless you want to suffer with a lopsided craft for the duration of the boat's life, be very careful about lining up the molds or building frames. Working outdoors, I like to use a 16-foot, two-by-twelve plank staked to the ground securely in a half-dozen places. Lacking such a plank, you can duplicate it with two two-by-sixes nailed level and parallel to the boatshop floor, or similarly staked out. Remember, you are building the boat upside-down.

My own temporary frames, built of one-by-fours, built securely and braced diagonally, have lasted through the building of seven boats, and they will be good for as many more boats if I ever get around to building them.

Build three true frames according to the accompanying diagrams. The first, or forward, frame is 38 inches high. You are building the boat upside-down, so this frame is 29 inches across what will be the bottom of the boat, and 48 inches across what will be the top side of the boat. The second, or middle, frame should also be 38 inches high, but it is 42 inches across the boat's bottom and 60 inches across the top. The third, or after, frame is 34 inches high, 34 inches across the boat's bottom, and 52 inches across the top. Be sure to cut out all slots indicated for keelson and chines. Secure the frames to the base, spacing them as indicated in the plans. Make sure that the centerlines are aligned exactly, and brace them both horizontally and diagonally.

The plywood sides, whether ⅜ inch or ½ inch, should be made with marine-grade plywood. Lacking that, you can use exterior "AC" grade, but always watch out for voids in the plywood. Two 4-by-8-foot sheets will be enough for the two sides. Since the sides are nowhere more than 22 inches high, rip the sheets exactly down the middle lengthwise with a "plywood" blade (if you're using a power saw) or a fine handsaw.

In skiff building, the frames must be braced securely. The whole shape of the craft depends on the precision of the frames.

Lay two of the 24-inch-wide pieces on the floor with the "A" side up (that is, the good side, which is to be outside). Just before you butt the pieces together, fill the joint with thick marine glue. You will need a foot-wide plywood butt-block at this joint. After covering the batten well with glue, clamp it securely on the underside, or "C" side, but make it stop just three inches short of what is to be the bottom edge, because you will be putting a chine in that three inches before you're through. Clamp the plywood pieces together.

Nail the butted edges to the butt-block first, working out from the center before the glue has set. Put two rows of nails, two inches apart, on each side of the joint. Scrape off the excess glue before it hardens. In the matter of fastenings, prices being what they are today, you're better off with ridged, bronze boat nails. They cost little more than galvanized screws, they're easier and quicker to use, and they will last longer. (Furthermore, too many of the galvanized or "hot-zinc-dipped" screws, which pass for galvanized screws, have the threads filled with the galvanizing material, which will rip the plywood.)

Transfer the boat's measurements onto one side, starting from a centerline and measuring to the top and bottom every two feet, as indicated in the plans. Put a limber batten on the points and scribe the lines. I like to use a saber saw to cut the curving lines, but a handsaw will do nearly as well. Some of the curves are pretty sharp for a circular saw, so, if you are using one, allow a little extra wood and plane off the surplus when the time comes. Cut off the forward end on the diagonal indicated, but leave the after end for later. After the first side is cut out, check it to see that the lines are smooth, without undue humps or hollows. Use it as a pattern for the other side (outside to outside, of course, so you don't end up with two port or two starboard sides).

Rabbeting out the solid stem and securing the sides to the stern are the most difficult tasks. I usually cut the rabbet and taper the stem with a hand-held power saw, but I don't advise this method for someone unfamiliar with the tool. Either have the stem rabbeted out on a saw table, or, if you want to do it the hard way, cut out the rabbet with a chisel. If it is a little deeper than the thickness of the plywood, don't worry; you can dress it after the sides are secured—just as long as both sides are even. Make sure the stem is a little extra-long, maybe as much as three feet too long. Allow the stem to extend a couple of inches above the sides when you are fastening them. (I say "above," but that means "below," since you are building the skiff upside down. To allow for the thickness of the bottom and the keelson, allow the stem to extend at least 1½ inches beyond the point where it joins the bottom of the skiff.)

(A word about the stem: a full four-by-four may be hard to come by. I have used hard pine, oak, and a driftwood piece of Honduras mahogany. It should be a clear-grained piece of some kind of hardwood, because the whole strength of the skiff will depend on the sides holding to the stem.)

Clamp the sides temporarily to the first two frames, with their edges about ½ inch above the frames and the bow far enough forward of the first frame so that when the sides are sprung in to fit the stem, the stem's inner (after) edge is centered and located 33 inches forward of the first frame. This joining is the most important

Planking the skiff.

in the whole construction. Make sure not only that the stem is centered but also that the sides are exactly level both fore and aft, the whole way.

It makes things easier if you nail a temporary extension board along the inside face of the stem piece. Nail the bottom end of the extension to the 16-foot base plank. This will be added support when you are ready to fasten on the bottom. After you have lined up the sides and the stem, coat the stem rabbet with glue and nail the sides securely, staggering the nails about two inches apart.

After the sides are nailed to the stem, clamp them to the third frame, again making sure that they line up exactly. You may have to ease the sides slightly at the first and second frames, but be sure you don't wring the stem out of line, and be sure the sides extend slightly, a half inch or so, above the frames.

Now for the chines. I am constantly amazed that so many stock boats have square chines, thus creating a water pocket between them and the sides. Use one-by-threes, either oak or mahogany, and bevel the edge that will be uppermost when the boat is turned right-side up, so that while it is three inches on the outside, it will be no more than 2½ inches on the inside. Spring the chines in the slots cut for them between the sides and the frames. Where each chine fits the stem, there will be a double bevel, so kerf it and kerf it again if you have to, so that it fits exactly. Once it is fitted (let the after end go; it can be tended to presently glue it well, clamp it securely to the side, and nail it, springing the after end down to fit the side ahead of yourself as you work from bow to stern. Nail it every three or four inches, staggering the nails top and bottom and letting the top edge of the chine extend slightly above the top edge of the side.

I do not like to make the transom out of plywood, because securing it to the sides means edge nailing. Instead, I use a double thickness of one-inch mahogany. But with the modern tendency toward bigger and more powerful outboards, and the consequent torque, the transom can't be reinforced too much. With furniture clamps, pull the stern in fair and true with the transom pieces, beveling carefully. I like to fit as carefully as I can, and then kerf with a fine-toothed saw. (You'll have to notch the chines to allow for the stern.) If I can, I make the grain run horizontally on the outer, or false, stern and make it run vertically on the inner stern. When the outer and inner pieces are glued together, there is a double thickness to nail into, and there is no need for end nailing. Stagger your nails an inch apart. Cut off the projecting ends of the sides flush with the stern.

Now you can plane the sides and chines where you had allowed them to project above the frames to take the bottom. Plane so that a six-inch board laid crosswise is in full contact everywhere. Don't skimp on this fairing, or the boat will leak forever.

If you are going to cross-board the bottom, you will first have to fit in the keelson, or inside lengthwise piece. At the forward end, start to taper a one-by-six about a foot back, taper and kerf it so it fits snugly against the stem and (at the very forward end) both sides. Put it in the slots cut into the frames, spring it to fit the stern, and secure it exactly midway between the sides. Clamp the keelson to each bottom board as the board is nailed. If you are using plywood for the bottom, it is better to add the keelson after the bottom, so you can turn the boat over, keep the fastening heads on the inside, and see just what you are doing.

Since the sides are spliced midway, I prefer to use a 4-by-12-foot piece of plywood for the bottom, either ½ or ¾ inch. Starting at the stern, and squared with it, clamp the plywood sheet with a ½-inch overlap astern, and the forward end square with the sides. The stern overlap is not necessary or even standard, but I find that it makes a slight "step," which helps the boat through the water. Scribe the bottom on both sides and cut. (Don't be afraid to be generous while you're scribing; it may mean more fairing, but it will give you a smoother job in the end.) After the bottom is cut out, lather the side edges and chines with glue, fit on the bottom and nail the bottom to the chines at two-inch intervals—that's one of the reasons you put the chines there.

Cut off the projecting end of the stem, flush with the sides and as far forward as the forward edge of the rabbets. A 12-foot piece of plywood on the bottom of a 14-foot skiff means that you will have to splice. You should have enough waste from the sides if you have used ½-inch plywood throughout; otherwise, you will need a little extra to fill in. Fit carefully, glue the butts and chines, and nail. You will fit in the batten when you turn the boat right-side up. If you have cross-planked the bottom, you can leave out the glue or use it as you wish, but chamfer each outside edge of the planks (not more than six-inch-wide planks, lest they warp), about three degrees if you're using a power saw, or ⅛ inch, to allow for caulking and swelling. A single strand of cotton caulking, rolled in, should be adequate if the wood is seasoned.

After the bottom is nailed, you can fasten your "keel" in place. It should be a piece of one-by-six oak, hard pine, or whatever, faired down to fit the stem (which can be cut off flush after the keel is in place) and tapering toward the stern, from about the third form, from six inches to four.

In the original plans for this boat, a skeg, tapering from one inch forward to four inches at the stern, was spiked to the keel. I have found it more practical to cut out of the keel a slot into which the skeg will fit tightly. So cut a slot in the keel lengthwise from the stern; make it 34 inches long and just wide enough to take the skeg. Cut the skeg so it increases gradually from one inch, plus the thickness of the keel, to four inches, plus the thickness of the keel. It is 34 inches long. Tack it in place so that it can be nailed through the bottom and into the keelson at the last minute. (Believe me, this skeg will make a world of difference in the way the boat handles, both when you are rowing and while you're fishing from it.)

Now you are ready to clean up the outside of the boat, ready to take it off the forms and turn it right-side up. Fair off the bottom, following the sheer of the sides. I usually use a jack plane for the rough work and finish off with a power disk sander, being careful not to cut into the plywood sides. If you haven't already made the stem flush with the bottom, do so at this point, and dress the stem flush with the sides.

After the boat is turned right-side up, if you have used plywood for the bottom, you will have a joint somewhat less than two feet back from the stem. I like to add a batten touching the chines, and running back from the stem far enough to lap the joint by at least six inches. Fill in this whole space, all the way back from the stem, and thus avoid a water pocket between the batten and the stem. You can use scrap plywood and even make filler and batten all one piece. Glue and nail.

Now, if you've used plywood for the bottom, the keelson may be cut out to fit over the batten forward, trimmed to fit stem and sides, and sprung in place so it fits at the stern snugly. Secure with fastenings that go all the way through into the keel. Since this piece will stiffen the boat, you may want to bolt through keel, bottom, and keelson. Use ¼-inch brass stove bolts in pairs, every two feet. (I'd prefer bronze bolts, but they aren't available locally.) Countersink the heads slightly on the outside, let the washers and nuts protrude slightly on top of the keelson.

I like to stiffen the sides by adding ribs every two feet for the length of the boat. Tapering from one inch at the rounded top, they run to two inches at the bottom, after they have been cut out carefully where they fit over the chines. Be sure they are cut off ¼ inch short of the bottom. Too many times, if they are left too long, they tend to pry off the bottom. Nail from the outside, except at the chines, where they are nailed from the inside.

To stiffen the boat and to provide risers for your thwarts, run a piece of one-by-two on each side, starting down about eight inches at

The skiff's interior after the boat has been removed from the frames. The port side is shown as installed, while the starboard side is shown after finishing.

the forward rib and down five inches at the stern. Fasten securely at each rib with two nails. Also for stiffening, and to protect the gunwales, attach rubrails of hardwood one-by-twos, appropriately rounded top and bottom and kerfed in to fit the stem exactly. Now cut off the stern, crowning it slightly in the center.

I have deliberately left out thwarts and rowlock pads, assuming that you will want to locate them for the convenience of whatever kind of fishing you plan to do. The thwarts can be secured to the risers, or simply fitted in place to be removed when necessary. The rowlock pads can be made of one-by-sixes, drilled to take rowlock sockets and bolted or nailed very securely to the sides.

The only job left is to secure the skeg. I like to use graduated spikes, driven right through the keel, starting with a 16-penny galvanized spike forward and boring holes and increasing the lengths of the spikes as I move aft.

Round off the top end of the stem and drill or not, as you see fit, for a mooring rope. Sand off the rough edges. Prime with any good paint thinned half and half with turpentine or paint thinner. (The whole secret in holding plywood in salt water is to keep it painted.) You can strike a waterline with a 16-foot piece of lattice or any limber batten, painting below with antifouling paint, and above the waterline with whatever color suits your fancy. If you paint the rub strakes a contrasting color, they will dress up your boat.

Since this is to be a work boat, you should have a bow chock port and starboard as close to the stem as possible (and big enough to take whatever size warp you plan to use), secured with long—very long—screws. A cleat should be secured similarly to each rail aft. I am likely to bolt on the cleats, right through the rubrails. It is better to have a slightly rough-looking job, rather than cleats that pull off just when the fishing is good.

You may want to add a grating to save the bottom in all kinds of fishing except eeling, where a grating can become a headache. Make the grating with one-by-threes of the cheapest lumber you can buy. Make it in three sections for easy removal, spacing the one-by-threes their own thickness apart and cutting them an inch short of the width of the boat where they bear.

Properly taken care of, dried out and repainted annually, there is no reason why this skiff shouldn't last for as long as you want to go fishing. (The first boat I built on these plans was done 22 years ago, and it is still in good condition. I built one out of plywood for our town seven years ago; it is used hard all year and still needs only a scraping and a coat of paint each spring.)

MATERIALS FOR A 14-FOOT PLYWOOD SKIFF

Sides: 2 pieces, 4′ x 8′ marine or exterior plywood, ⅜ or ½ inch.
Bottom: 1 piece, 4′ x 12′ marine or exterior plywood (or 2 more 4′ x 8′ pieces). If you use ½-inch plywood throughout, you will have enough scraps to finish off the bottom and also make your battens. If you use ¾-inch plywood on the bottom, you'll need another piece of ¾-inch plywood, 4′ x 4′, but you can also make your thwarts out of this.
Stem: 1 piece, 4″ x 4″ (full), 30 inches long, hard pine or oak.
Ribs: 2 pieces, 1″ x 2″, 10 feet long, oak or mahogany.
Rubrails and risers: 4 pieces, 1″ x 2″, 16 feet long, oak or mahogany.
Chines: 2 pieces, 1″ x 3″, 14 feet long, oak or mahogany.
Keel and keelson: 2 pieces, 1″ x 6″, 14 feet long, oak or mahogany.
Stern (double thickness): 1 piece, 1″ x 8″, 16 feet long, oak or mahogany.
Frames and braces: Scrap 1″ x 4″.
Fastenings: Glue, bronze nails, cleats, chocks, rowlock sockets, and 14 ¼-inch brass stove bolts for the keel.

GLOSSARY

ADDUCTOR MUSCLE: the tissue that connects the two shells of a bivalve. In scallops, the only part a Cape Codder eats.

ALEWIVES: see HERRING.

BIRD'S NEST: a snarl of line caused when the spool of a free-spool reel overruns itself, or when loops of line slip off the face of a spinning reel.

BLUEFISH: *Pomatomus salatrix*; oceanic, chiefly warm-water fish, seldom but occasionally over 20 pounds. Known chiefly for its savage teeth and voracious nature.

BULLRAKE: also called "longrake." A toothed tool followed by some sort of basket, secured at the end of a long pole 10 to 40 or more feet in length. Used to gather quahogs (and, in some waters, oysters) in water too deep for other methods.

BYSSI: the long, tough beard, hair, or filaments by which some shellfish, particularly blue mussels, are attached together.

CAR: a crate or floating pen used to hold live fish, crustaceans, or shellfish.

CHAFING GEAR: rope ends or similar substance tied at the knots on the bottom of an otter trawl to prevent wear. Occasionally a whole cowhide was used for this purpose. Also, any protective wrapping.

CHAMFER: to cut a groove, furrow, or bevel.

CHRISTMAS TREE RIG: also called "coat-hanger rig." A multiple lure used in trolling for striped bass and bluefish.

CHUB: *Fundulus heteroclitus*; also called "mummichog" and "killifish." A stout-bodied minnow that seldom grows longer than three or four inches. It abounds in tidal creeks in salt marshes, but lives in brackish water at the mouths of streams and estuaries.

COD END: the heavy netting pocket at the after end of an otter trawl. Usually fitted with a pucker string, which, when untied, releases the fish on deck.

CODFISH: *Gadua callarias;* also called "cod" and "rock cod." The staple fish in northern waters, distinguished by its barbel, three dorsal fins, and strong lateral line. It occurs in many colors, from almost black, through the browns and reds, to green, and the belly is usually whitish. In the surf, codfish usually range from six to 12 pounds. Offshore, 50- to 60-pounders are not uncommon.

COKE FORK: a long-tined, short-handled tool originally designed to shovel coke or crushed stone, but adopted by commercial mussel fishermen.

CULLING BOARD: a platform athwartships and usually resting on the rails of a skiff. The drag load is dumped on the board and the trash is pushed overboard as the shellfish are collected.

DINOFLAGELLATE: any of an order of chiefly marine planktonic forms, usually solitary, plantlike.

DIP NET: a pocket of twine mesh, usually rigged on a metal hoop, which is secured to the end of a short pole. Used for catching fish, crabs, or shellfish.

EELGRASS: *Zostera marina;* a submerged marine plant with long, narrow leaves; a true grass; a hiding place for baitfish and shellfish spawn; and a headache for the troller when it is drifting on the surface.

FOOT ROPE: the rope at the foot or bottom of the leading edge of an otter trawl, usually weighted with lead or chain.

FAKE DOWN: to coil down so it is free to run.

FIDDLER CRAB: *Uca pugnax*; named after the male's huge single claw. Also called "army crab" and "soldier crab." A small (two-inch) burrowing crab that digs tunnels up to three feet deep. If the male loses its right claw, it can grow a left one; the female has no such giant claw.

FLOUNDER: almost any of the flatfishes, hereabouts usually *Pseudopleuronectes americanus,* "winter flounder," "blackback," or "Georges Bank flounder"; a small-mouthed, right-handed flounder without teeth. Has been known to grow to eight pounds, but alongshore seldom runs over five pounds.

FLUKE: *Paralichthys dentatus,* "summer flounder," "plaice," or "plaicefish"; a left-handed, large-mouthed flounder with teeth. Seldom grows larger than 15 pounds in the Cape Cod area.

FYKE: stationary, and therefore more or less permanent, eel trap, usually set in a salt marsh.

GURRY: fishing offal, waste; guts-heads-slime-and-scales.

HANDLINER: a fisherman who uses a handline or drop line; a boat used by such a fisherman.

HEADBOATS: derisively and rather obliquely called "plunger boats" by charterboatmen; also called "party boats." The distinction is chiefly that the charterboats usually charter the whole boat to one party, whereas in headboats, the passengers or fishermen pay only so much per head or person.

HEAD ROPE: the rope at the head or top of the leading edge of an otter trawl, to which is usually secured some sort of flotation device—net corks, glass or metal floats, etc.

HERMIT CRAB: *Pagurus bernhardus*; small, soft-bodied crabs found near low tide lines. They seek out shells to use as "houses," abandoning one shell and seeking a larger one as they grow.

HERRING: on Cape Cod, "herring" usually refers to alewives (*Pomolobus pseudoharengus*), although seven species of herring occur along this coast. The alewife, distinguished by the greater depth of its body, is common on Cape Cod only in spring, when it runs up the creeks to spawn in fresh water.

HIGH LINE: top dog; a fisherman who customarily or on any particular day catches the most fish.

KERF: a slit the width of a saw cut. To kerf means to carve a very narrow slit with a saw.

MARICULTURE: farming or cultivating the sea, usually interchangeable with "aquaculture."

MERMAID'S HAIR: a blue-green alga living on mud, rocks, and pilings in shallow water. The simple, unbranched filaments are curled and matted.

MOONSNAIL: *Polinices heros* or *Polinices duplicata.* The former, more commonly called "northern moon shell," ranges from Boston northward, whereas the latter ranges from Cape Cod southward. (Their territories overlap somewhat.) Also known locally as "cockles" and "round wrinkles." In certain areas, the moonsnail is the very worst shellfish predator, burrowing as much as six inches after soft-shell clams.

OTTER TRAWL: a drag or pocket of netting made to be towed along the bottom of the ocean to catch fish; said to be named thus because it "fishes like an otter."

PARLOR: the after chamber of a two-part trap for fish, lobsters, or crabs.

PLAICE: see FLUKE.

PLANING BOARD: a piece of plywood or sheet metal secured crossways to a scallop drag at such an angle as to plane or divert the water over the drag. Properly rigged, it will cause the water to rush over the drag, thus creating a sort of vacuum to lift the scallops off the bottom and into the drag.

PUCKER STRING: a rope rove or threaded through rings secured to the after end of a cod end, pulled tight and tied while fishing, untied and spread while dumping fish on board.

PYRAMID SINKER: a tapered sinker rigged to slide on the surfcaster's line, and designed to dig into the bottom.

QUAHOG: *Mercenaria mercenaria*; also called "round clam," "hard clam," "littleneck clam," and "cherrystone clam." A round, hard-shelled bivalve, deeper than it is thick. Ranges from some sheltered bays and harbors in Maine to Florida.

RAZOR CLAM: *Ensis directus;* called "razor clam" because it looks like an old-fashioned straight razor. It occurs all along the Atlantic Coast in mud or sand near low water.

ROCKWEED: a coarse, brown seaweed that attaches itself to rocks or spiles, with air bladders along its main stem.

SEA LETTUCE: largest of the green algae, up to three feet long. It is sheetlike or ribbonlike.

SEA WORM: also called "sand worm." Any of several large, burrowing bait worms, especially of the genus *Nereis.*

SCHOOLIE: commonly understood to mean small striped bass (since they travel in schools).

SCULL: to propel a boat forward by the use of one oar thwartwise at the stern.

SQUETEAGUE: see WEAKFISH.

SQUID: a long, 10-armed cephalopod, usually divided locally into "winter squid," ranging from 12 to 20 inches long, with a "pen" or internal backbone, which appears to be made of lucite or celluloid, and "summer squid," smaller and without the "pen." It swims by jetting out water, and protects itself by ejecting a cloud of "ink," cloudy, black fluid. It is used chiefly as bait locally, although there is a strong movement afoot to use it as food.

STILES: the poles or handles, seldom more than 16 feet long, with which tongs are worked.

STRIPED BASS: *Roccus saxatilis;* also known as "striper," and in more southern waters as "rock" or "rockfish." *The* northern sportfish; fairly frequently reaches 50 pounds, seldom over 60.

TAUTOG: *Tautoga onitis;* also called "blackfish" and "whitechin"; of the wrasse family, it ranges in color from almost black, more commonly brown, to green, usually mottled. It feeds on crabs and barnacles and juvenile shellfish.

TREADING: feeling for shellfish with one's feet, usually barefoot in the mud.

WARP: a rope (or cable, in which case it is usually called "wire") extending from boat to anchor (anchor warp) or to a drag (drag warp). To "warp" is to pull, by rope or cable, a boat to a fixed object such as a dock.

WEAKFISH: *Cynoscion regalis;* also known as "squeteague" and "sea trout." Gone from Cape Cod waters since the late 1920s, it seems to be on the increase. Very much like a freshwater trout in appearance, it is reputed to have reached 30 pounds, but a 17-pounder is nothing to be ashamed of. Called weakfish not because it is weak but because its mouth is very tender. A summer or warm-water fish.

WINGS (of an otter trawl): the relatively narrow bands of netting or twine extending from the boards or doors to the pocket itself.